Bee Wisdom
Teachings from the Hive

Sandira Belia

Bee Wisdom
Teachings from the Hive

Translated by Linnéa Rowlatt

Northern Bee Books

Bee Wisdom Teachings from the Hive
Copyright © Sandira Belia
Translation Copyright © Linnéa Rowlatt

Illustrations by Sandira Belia
Photographs of the paintings: Simon du Vinage

Published 2021 by
Northern Bee Books, Scout Bottom Farm,
Mytholmroyd, West Yorkshire HX7 5JS
Tel: 01422 882751 Fax: 01422 886157
www.northernbeebooks.co.uk

ISBN 978-1-914934-18-6

Design and artwork DM Design and Print

To my mother,
to my grandmother,
to Mother Earth.

TABLE OF CONTENTS

PREFACE

Morning and evening I have a ritual that opens and closes my day: I listen to the bees. We share a common wall off my studio. When I waken and then again when I am ready for sleep, I pad in barefoot and put my ear up to the hive wall to hear the bees singing. I let the sound enter me, eyes closed and quiet. I listen with each ear, so both sides of my brain absorb the song and permeate my consciousness. Listening deeply, the sound enters my heart. If it is evening, I hope to dream about bees and sometimes I do.

I spent time yesterday with my beloved hives. I have many different hives scattered about our farm, each happily habituated or ready for a swarm to find and claim. Because I live in the rainy Pacific Northwest, each hive has a rooftop hat to keep them dry in the rainy seasons. Every hive is different, each with its own unique character. Together the hives sprawl like a village across the field.

"What kinds of hives?" you might ask. Probably none that most beekeepers would expect. No thin-walled square wooden boxes here! Instead we have beehives patterned after bee homes made in far away countries centuries ago. We have novel hives designed by people who think bees' needs are the priority. We have clever hives made by good friends and filled with love. And a few hives designed by my intuition.

I visited each colony and stood or sat alongside the front entrance for awhile observing everyone's tasks and demeanor. My to-do list for the day included trimming the grass away from the front of each entrance, neatening up my spare equipment pile, and reveling in each colony's progress.

As I walked the paths between hives, I thought of Sandira's magnificent book and how generous she is with her creativity and knowledge. I was thinking of you, dear reader, and what delight you will have sharing the joyful bees' world view that encourages them (and us!) to live in harmony with Nature. It brings me great joy that so many of us are listening and learning from the bees, choosing to see and understand the world from their perspective.

We need more bee-centric thinking. We need to look to the past and find methods from times when bees were more robust. We need principles that respect bees into the future. We need courage to step away from today's common practices and prioritize compassionate methods that put environmental awareness at the forefront of all our decisions about tending bees. We all will benefit from a larger, longterm view of the consequences of what we do with bees. Over time, this knowledge will help bees become healthier and delightfully happy.

As I was turning away from one hive, I caught sight of a dangling dead bee under the entrance. Sure enough, a spider had setup shop under the hive and was doing a thriving business in bee-steaks.

I want to share a lesson in observation that I frequently use when the bees have a differing opinion:

Question everything and trust your bees.

They understand their needs better than we do. Humans can be terribly arrogant when we insist we know best. Support what the bees want.

So I got out my spider-web-clearing brush. I purchased this yellow nylon brush to sweep bees off comb, a 'task' I no longer have need for, because I don't move combs and I let bees stand wherever they want.

All my bees despise this brush. I bought it because it was advertised as a gentle way to move bees off comb without disturbing them. I wondered why the bees would have such strong feeling about this soft-haired bee sweeper.

I saw them be aggressive to the brush a few times so I stopped using it and I put together a two part hypothesis:

(1) Bees don't like to be swept.

(2) I bet I surprised a bee when I was sweeping bees the comb and, in her startled state, she threw a stinger in. Another bee smelled the venom pheromone and stung the brush hairs, too, and now that brush probably stinks of danger and upset.

In response, I stopped doing a thing that bothered them.

When you see something you don't understand, don't assume human superiority. Instead step back and watch. Over time you will amass a thousand hours of questions and, hopefully, a few minutes of knowledge.

The bees don't want that brush to touch them so I no longer use it near them. I've relegated that nasty brush to a new function, spider web cleaning. I went under-over-and-around every hive box, table, chair, shelf, corner and window frame, brushing away webs and — soft-hearted sot that I am — moving spiders outside into the field.

I got sidelined at one point when I discovered a praying mantis chrysalis under a window sill and, oh happy day, the miniscule baby mantises were hatching at that very moment. Each one stretched its scarecrow limbs from the pod, scooted with legs akimbo across the wall, then squeezed through the gaps into the thick grass of the great outdoors. Fifteen mantis birthdays right before my own eyes.

And then it was dusk. I hadn't cut the grass, I hadn't reorganized my equipment. I had visited most of the hives for 2.5 hours, but how could I call that work time when not one task had been completed?

That's how it goes at our farm. I get an idea of what I'll do today, then on my way to doing it, I find something else that needs attention. The well-structured intentions I started my day with often aren't what I am doing as the sun sets. Nonetheless, this is all part of working toward my ten-thousand-hour bee education, upon which when I complete it, I hope to actually know something about the nature of bees.

Even with twenty long bee years behind me, I am still a beginner who is surprised every day by what I see in the bee yard.

This is what a life shared with bees looks like. The Bee Deva's rich song enlivens layers of harmonies. The visions shared reveal a sacred path where we are One. I am joined together with Sandira and with you through the sweet-tongued family of bees.

Bless us all.
Jacqueline Freeman
Author of *Song of Increase:*
Listening to the Wisdom of Honeybees for Kinder Beekeeping and a Better World.

TRANSLATOR'S NOTE

I am deeply honoured to have participated in bringing Sandira's book to the English-reading audience. Frequently during the last year, I felt the Bee Deva hovering nearby, supporting my efforts to reach the spirit of the message and to translate it appropriately. This accuracy was the goal of my work and I made translating choices which foster it, rather than adhering to a single method or approach.

Linnéa Rowlatt, PhD
Whitehorse, Canada
September 2020

ACKNOWLEDGMENTS

Thank you to all the BeeShippers for making me love my work so much and for our fabulous shared moments.

Infinite gratitude to my dear Annelieke for supporting me through all the ups and downs of my 'fibro years' and, through her love of bees and her faith in the essence of human nature, for having accompanied and co-created the energetic framework of this work.

Deep gratitude to Linnéa for her perceptive feedback and her dedication to completing this beautiful translation of the original manuscript.

Thank you, Long Grass, for your elegant translation of my poems.

Thank you to my teaching hives and to the Bee Deva for adopting me and taking me under their wings, for showering me with their wisdom, and for this gift offered to the world.

INTRODUCTION

The world is curious about the bee.

A little pollinating fairy with six legs, humble and discreet, her fervour fascinates and her wisdom intrigues. She does not have the silky fur of a baby seal nor the mischievous smile of a dolphin to make us melt and yet, she touches our hearts with a profundity that is difficult to describe. Her mysterious song sets off a particular resonance within the human being which extends beyond the mental and the emotional, a sort of soul resonance... What is this irresistible charm that she emanates? Where does it come from, this magic which seduces us?

The media regularly draws our attention to threats facing the bees' well-being, awakening compassion and curiosity about them. A good portion of humanity is beginning to be aware that we are only part of a Whole, that our planet is a living being of whom we are the cells, and that all are interconnected with one another. The bee is one of the emblems of this enlarged perspective. We have come to realise that we are not individuals separated from each other and that each thought, each word, and each action has a repercussion on the *ensemble*. Thus, with our daily choices, we have the power to change the world...

This realisation is frightening on many levels, since it brings us face to face with our responsibility as members of the terrestrial family, and that brings certain old habits into question. At the same time, feelings of fullness, joy and hope emerge when one perceives the immensity of the potential that this represents. To learn to unite our strengths, not only among humans but also with the animals, plants, and Devas,[1] opens the doors to co-create a veritable earthly paradise: *a planetary garden*.

The bee recalls us to our Soul. One wing in the wild world and one in the domestic, she offers herself as a creator of relationships, a guide and a teacher of the complex and intriguing mysteries of the Universe. The bee is a companion, a counsellor, an ally. She unites the personal and the universal.

Throughout our history, numerous traditions consider the bee a messenger of the Gods, recognising her characteristics as a mediator and carrier of information between worlds. Engraved on tombs, she has long been honoured as a companion of souls during their passage from life to death, and, invoked during birth, from death towards life as well. For a long time,

---

1   In Sanskrit, *Deva* means 'light being'. I choose to bestow the feminine gender upon her throughout this work.

the bee has been enveloped in a cloak of beliefs accepted as established truth without having been truly understood. Only a few contemplative philosophers studied her teachings, along with certain lodges confined within Mystery Schools. Today, the time is ripe for this knowledge to be revealed to a larger and more varied body of truth seekers eager for learning and wisdom.

The bee *knows* and she awaits us.

The Call

I was born in Brittany, France, to a mother with a deep passion for bees, an apiculturalist, and I bathed in bee song from my earliest years. When barely as high as three apples, I was already approaching swarms in search of a new home without a bee veil and without fear. My favourite smell was my mother's scent during the harvest: a mixture of smoke, propolis, and the sweat of her happy labour. Golden honey sang of summery richness, and I delighted in watching it extracted from the comb and poured into jars.

Growing up, I never dreamed I would become a beekeeper, maybe because my inner rebel refused to be reduced to 'being like my mother'. Nevertheless, in 2012, I was offered the opportunity to take responsibility for the bees of the community of Tamera, where I was then living.[2] A spontaneous wave of shivers ran over my skin, an inner barometer conveying my cellular resonance with the proposal. A joyous "Yes!" erupted from my womb, to which I straightaway added, "But don't count on me to furnish the community with honey!" Indeed, this visceral call was to something else, to a new relationship with this mysterious little being...

Even though it might seem odd, the bees speak to me. I listen and speak to them in turn. The conversations which take place are sometimes so astonishing! Their messages are stuffed with a variety of information that I have not heard, sometimes, anywhere else. I refer to this being with whom I converse as the 'Bee Deva'.

The Bee Deva is a group Soul, a collective consciousness. She is not identified with a specific colony and even though it is easier for me to contact her when I am in proximity to a hive, this is not an indispensable condition. Our communications take place through a variety of means which are physical or subtle by degrees. Sometimes the Bee Deva visits me in an

---

2   A centre of research and education for peace, Tamera is an intentional community holding around one hundred and twenty permanent members (as of 2018), situated in the heart of the Alentejo in Portugal. Website: www.tamera.org

energetic form, invisible to my physical eyes but clearly perceptible and tangible to my subtle senses. At other times, she sends a messenger to me: a bee who flies around me buzzing at an unusual frequency or who rests on my neck and murmurs into my ear. It has been for several years now that I have been receiving this information with respect, fascination, and humility, and it is with great joy that I share a few selected pieces of it here, throughout this work. This book is a crossroads, a meeting point between ecology, art, science, and philosophy. Just like the Bee Deva, its role is to bridge the worlds.

The weave of this book is based on my personal experience. The more my intimacy with the Bee Deva deepens, the more she reveals her knowledge. The information she transmits comes to me in different forms. Sometimes it arrives as an idea formulated to greater or smaller degrees, and sometimes the messages are visual – in colours, forms, or impressions. I collect and compile this information using my antennae, and then my conscious mind applies itself to transcribing the message into words, sentences, and paragraphs. I feel somewhat like an interpreter who translates passages from the Book of Nature into human language. As the material journeys through the library of my history, its expression is forged and tinted by the gouges and brushes of my inner landscape.

Therefore, dear Readers, please do not *believe* me! Do not take anything which I tell you here as true without running it past the acumen of your own intuition. How does it feel? Receive, chew over, savour the images and the words I put forward - always holding to the spirit that all truth is relative, and that there is nothing like direct experience for validating information. While reading this book, I invite you to invoke the presence of the Bee Deva for yourself, and to hear her messages using your own words, those which resonate most powerfully within your heart. Equally, dare to explore, to test, and to enrich this research, to ask new questions and to share your answers. And if the information presented in this book seems improbable to you, out of balance or overblown, you may choose to read the book as science fiction or a fairy tale...

Bon voyage!

Sandira Belia,
Wednesday, 10 October 2018
Vale Bacias, Alentejo, Portugal

1
FROM SOUL TO SOUL

Human soul, human friend,
Our roads commingle and carry us beyond,
Listen, feel, and trust me... I love you!

A Bath of Love and Abundance

Sunday, 28 May

To take a bee bath, I settle down on the mattress positioned beneath *Prâna*.[1] I insert the two stethoscope tips into holes drilled for this purpose on the back of the hive, and place the earbuds in my ears.

The song of the hive floods me. I find myself in a cathedral, bystander to a concert of thousands of choristers in symphony. The bees offer their song of abundance, the best remedy to dispel all my worries... I allow myself to melt in this chorus of extraordinary harmonics. Scented fragrances reach my nostrils, the nectar abounds. Calmly fanned, skilfully alchemized, it is transformed into gold...

Tenderly, from the bosom of the choir, the Bee Deva murmurs:

Soul friend, human Soul,
Our roads carry us beyond reality,
Listen, feel, and trust me... I love you.

My brainwaves slow down and I fall into a semi-doze. Inhale, the colony enters into me... Exhale, I dissolve in her... My body soaks in her generous song; I am intoxicated with an invisible honey. The linearity of time collapses, the fragmentation of space fades away, and the boundaries of my self evaporate...

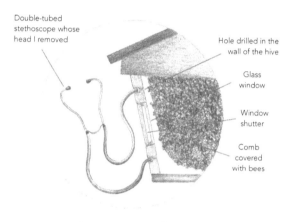

Double-tubed stethoscope whose head I removed

Hole drilled in the wall of the hive

Glass window

Window shutter

Comb covered with bees

Cross-section of the hive

---

1   *Prâna* is a colony installed in a horizontal *topbar* hive 'on legs'. I put a pallet beneath the hive, on which a mattress may be placed to take a 'bee bath'.

The Bee Deva is multidimensional. Her knowledge of the architecture of the physical and subtle worlds is extraordinary. This knowledge makes her a shaman of the animal world. Healer, muse and teacher together, she offers herself as a guide and mediator, humble and devoted.

The Healer Bee

For several years now, life has been throwing a challenge at me - something both fascinating and exhausting, a kind of enigma to solve, an apprenticeship to navigate: I am confronted with a panel of physical and psychological symptoms currently known as 'fibromyalgia'. Chronic muscular and joint pain coupled with a dysfunctional and hypersensitive digestion have led me to a drastic loss of weight and reduced capacity in my legs. The path of healing is rough, dotted with potholes and bumps, tantrums and revelations, frustration and gratitude. Although uncomfortable, the 'dis-ease' is a fascinating voyage, a school of life which offers in-depth teachings about the mysteries of human nature. Upon this meandering road, I have come across allies of all sorts, more or less luminous or tenebrous. Whether from their support or the challenges they pose, all of them invite me to advance. Among them, I met the Bee Deva. She presented herself to me as a companion, a counsellor in whom I could confide my doubts and questions. She offered me her remedies, both physical – such as honey, bee bread, and venom – and energetic.

The true healer is not a saviour who removes the victim's misfortune with a stroke of the magic wand, the Bee Deva tells me. *For healing to be effective and enduring, the role of the healer is to accompany the subject in their process of self-healing. The Healer opens this space of trust in which the Being in transformation can access their own healing potential. Healers assist the body to remember its true essence.*

Thusly does the Bee Deva propose her guidance to our own inner healer, in order that we may become agents in our own healing. And this cannot be done without understanding the teachings that are revealed all along the journey. *The ultimate healing*, adds the Deva, *is to recognise the extent of one's inner power.*

Wisdom and healing complement and support each other like the double helixes of a strand of DNA. Each step on the path of healing carries a teaching, and each new insight is a healing.

The Bee Muse

Every artist has their muse and the bee is mine. She is a teller of stories, inspiring me to give them form through painting, song, and writing. The artist is an interpreter who renders visible the invisible. Art provokes, art questions. It asks us to look, to listen, and to feel beyond the forms. The inspiration which guides art is steeped in the spheres of the impalpable, and reveals its mysteries in colours, in games of light and depth.

Each painting in the series presented in this book was an adventure in its own right, a true journeyman's labour. The canvas is like a window into which I dive deeper each day of its revelation under the bristles of my brush. To begin with, the idea knocks on my door like a sweet song. She innocently shows herself; I sniff around her and hum her song, which the Deva teaches me. Then the idea descends further and the image is formed: liquid, at first, then viscous, and finally it crystallises, like nectar matured into honey... The picture is born and refined day after day. Often, I have to hide it away for a while, sometimes seal the cell for several months while the gestation continues, until it dares to come out of the hive...

The Teaching Bee

I have been communicating with the Bee Deva for several years now, and as our contacts have grown, our conversations have become increasingly clear and precise. At the start, I didn't really understand to whom – or to what – I was addressing myself. In my numerous notebooks relating to our exchanges, the Bee Deva frequently changed name and the semantic field of her messages has greatly evolved with the passage of time.

The Bee Deva does not speak a human language. When we communicate, the conversation takes place inside my brain and uses my inner semantic and cultural library to build the sentences of our exchanges. Sometimes the phrases are formed in my unconscious and appear directly before me, in French or in English (the first being my mother tongue and the second, the one that I use in daily life). Most of the time, I receive the concepts and content of the messages through different channels of perception in the form of patterns and visual, auditory, or kinesthetic sensations that I transcribe into human language. It often happens that the Deva orients me towards a specific book or documentary film for the purpose of enriching my cultural and scientific store of knowledge – which then permits me to illustrate her subjects in the most appropriate manner. She encourages me to formulate clear questions. *Every question contains its answer*, she says, *when the question takes form in the mind, its vibratory field is activated, attracting the revelation of its answer.*

In order that her message may be presented as clearly as possible, we have agreed on the following as definitions for the words and expressions which we use in this book.

The Voice of the Bee

"Humans have learned to think that it is matter which defines space. We invite you to reverse this perspective and envision that it is Space which defines matter. Space is the original *Source*[2] from which all form emerges and to which all form returns. Source is not a definable place. Eternal and infinite, it is at the heart of everything.

The *Soul* is the intermediary between Source and form; the material world is its field of experimentation. As though on a screen or onstage, the world of form is the projection of a reality within which the Soul experiences the joy of incarnation and of co-creativity. Soul exists without the body, but the body cannot exist without Soul. The *body* is the physical temple which conveys the jewels of the Soul. This visible material envelope is but one tiny part of the Being. Quantum physics now reveals that what you commonly call 'matter' is, in reality, 99.9999999 % made of Space. Even though it is commonly called a 'Void', Space is full; it is a vast field of electro-magnetic vibration, an infinite sea of energy.

---

2   The word 'Source' is interchangeable with the following words or expressions: Universe, Space, Void, Unified Field, Zero Point Field, Ether, Original force of Creation, Primordial energy of Life, Great Spirit, God, or any other appellation of your taste.

The *Bee Deva* is the collective Soul of the bee. Between the collective Soul and the physical body of each colony is the *Hive Being*.[3] Each hive is the physical manifestation of her own Being. The Hive Beings are governed by the Bee Deva. Each Hive Being carries her own history, which matures and is enriched from incarnation to incarnation. Although endowed with her own personality, the Hive Being is devoid of ego; she is not conscious of herself as an 'I'.[4] All her attention and energy is purely devoted to the service of All.

And what is *consciousness*, in all this? Consciousness is a function that allows Soul to know itself. In essence, the Soul is perfect and has no need of form, but the experience of material existence allows for an expansion of the field of consciousness. Each experience that we have while in form (whether we are a bee, a pebble, a tree, or a human being) constantly reverberates in the Soul and, as a consequence, in Source. In the mirror of physical experience, the Soul learns to re-cognise itself.

The human Soul has the unique capacity for incarnating with a high level of individual consciousness; you are capable of knowing yourself in a mirror and defining yourself as 'me'. The level of individual consciousness in the Bee is much less, but she is endowed with an extended collective consciousness. Nowadays, the human being has a tendency to learn about the world from the limited perspective of the physical body. The Bee Deva invites you to live and to perceive your life from the level of Soul. Soul-to-Soul communication offers a much wider and deeper intimacy. The Soul's world is free from the linearity of space and time. The myth of separation is obsolete. This is where genuine contact can occur."

---

3   Also called the Bee Spirit, the Group Soul of the bees, the Collective Consciousness of bees, the Bee Being, the morphic field of the bee, the collective memory of the bee...

4   The term *Hive Being* is translated from the French original 'Génie de la ruche' (*translator*).

Devic Languages

Thursday, 13 December

Yesterday morning, I met Andrej, a blacksmith from Germany who is passionate about his work. I, who have always felt a certain resistance towards the metal element, listen to him speak with fascination. Through his words, it is the Deva of Metal I hear expressing herself. Andrej speaks about the furnace of his forge as a womb which carries and gives birth to his work. He describes the beauty of the patterns which appear on the surface of a forged blade as it cools, and with a sweet smile, he confides to me that as he hammers the iron, he is actually caressing and sculpting it with the tenderness of a lover. "A deep healing must happen between humanity and Metal," he tells me. "The blade is a noble tool whose role is to serve the world." Our conversation takes place a few metres from the hive *Prâna*, and while Andrej is speaking, several bees alight gently upon him – something which they do not do very often. Obviously, they feel that Andrej is expressing himself entirely from his heart.

Devas are everywhere: at the heart of every animal, every plant, and every object in our daily lives. They constantly interact with our thoughts and our behaviour. As we become aware of their language, our relationship with the world changes dramatically. The Devas invite us to play with them, like children do...

Receptivity to their language of metaphor, symbol, and synchronicity requires a certain open-mindedness. Our Western lifestyle has accustomed us (most of us) to maintaining a high level of *Beta*-wave brain frequencies; *Beta* brainwaves characterise so-called 'ordinary' thoughts, those which analyse, reflect, and discuss. This incessant flow of thoughts encumbers our spirit and leaves us unreceptive to guidance from the Devas.

Some practices, such as shamanic voyaging, meditation, chant, or dance, encourage the development of other ranges of brain frequency, more suitable for subtle perception and receptivity to Devic messages (notably *Theta* waves). The encumbrance of the mental space recedes, opening the way to new dimensions of reality. The Bee Deva names this space the *hummm*.

To live in the *hummm* on a daily basis involves undoing the analytical mind's addiction to judgment, to pondering or commenting on every detail of our life, in favour of developing a new connection to the world, a relationship from Soul to Soul. The *hummm* invites trust and simplicity, and a surrender to allies whose understanding surpasses the capacity of the rational mind.

A Millennial Contract

From the onset of our conversations, the Bee Deva has been telling me that humanity and the bee, long before their arrival on Earth, signed a sort of *contract* that connects us. It was difficult for me to grasp this idea, as, in my mind, it inferred the notion that this made one dependent upon the other. That did not correspond to my view of the Universe, where I understood each being as interdependent, free and responsible for their own evolution. I came to realise that this concept of 'contract' was not accompanied by same constraints and obligations as generally implied by human contracts. *Au contraire*, this concept valued both Humanity and the Bee for their respective potential. Here is what my devic friend had to say about it:

The Voice of the Bee

"All the beings which make up Earth's Nature are partners in a great contract, a co-creative communal project, within which each one has their particular qualities, their specific function and their responsibilities to assume. At the heart of this participatory network, certain species have developed specific connections between themselves that are stronger than others. These are sub-contracts in the contract, so to speak. This is the case between bees and humanity. A millennial, sacred partnership connects us.

Like the human being, the Bee Deva is of extraterrestrial origin. We chose to incarnate on earth in the form of a social insect organised in colonies. Our biophysical and social structures offer particularly interesting attributes to our partnership. As well, we are repositories of deep wisdom about the natural cycles, into which we are able to initiate you. On your side, you humans are endowed with exceptional mental, emotional, and physical potential, which you can learn to develop and to place in service of planetary cooperation. Humans and bees can unite their respective strengths so as to participate in the expansion of the consciousness body of Gaia, our Mother Earth.

Over the last few thousand years, the majority of human beings have forgotten this contract. The intensive mechanisation of your agriculture, allied to a mercantile and productivist thinking, has reduced our relationship to that of master and slave. But forgotten is not annulled. The contract is still there, signed and engraved within the memories of the world, ready to resurface now that humanity has grown. We await you with patience and benevolence, for the bee does not know bitterness. It is with a thrilling joy that we find numerous human beings awakening, deciding to listen and to offer their talents to the world. Marvellous, the time is ripe…"

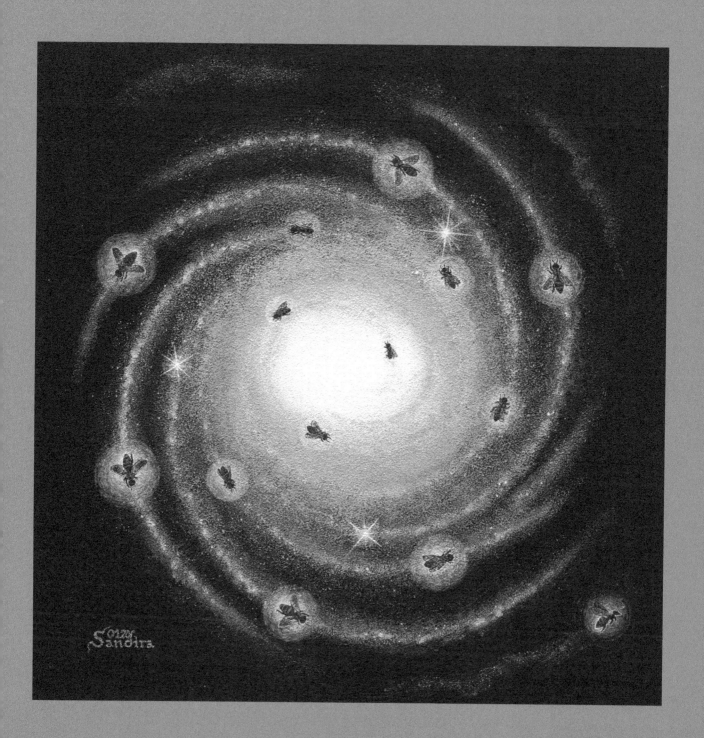

2

A SUPERORGANISM

Are you one, or are you many?
Singular body of multiple cells,
String of sparkling, fractal pearls.

A Human Hive

Friday, 28 September

Doorn Forest, heart of the Netherlands. I am participating in the first international conference *Learning from the Bees*, where more than three hundred bee lovers, artists, researchers, philosophers, and beekeepers from around the world are assembled. The atmosphere is buzzing and joyous. Participants have smiles on their lips and diamonds in their eyes. We are aware that this meeting marks a considerable step towards a relationship between Humanity and Bee that is based on listening and respect.

This evening, I join with some forty people in creating a 'bee-constellation', a workshop guided by Michael, Cheyanna, and Elena.[1] When I enter the room, I don't anticipate that the experience I am about to have will transform my fundamental perception of the world.

In the course of some preparatory exercises, we are invited to melt into the largest question in the world: "Who am I?" Then we are invited to consciously penetrate an oval space outlined on the floor by a rope: *the hive*. The instruction is simple: "Allow yourselves to be guided by the question 'Who am I?' without thinking about it rationally. Move, hum, interact... until the gong sounds." After some twenty minutes, our human hive freezes at the sound of the gong.

We observe and analyse the constellation designed collectively, along with the experience as lived. For me, this 'screen shot' is a moment of great revelation. The image which we formed – a reflection of our collective morphic field – is a combination of various geometric patterns: spirals, concentric circles, and lemniscates.[2] The position, posture, and dynamic of each individual reveals certain qualities of their Being and their relationship to the world at that precise moment of their existence. This 'freeze-frame' image renders visible a deep-seated universal mathematic of our existence. I am touched by its beauty and its relevance.

---

1   Michael Joshin Thiele (*Apis Arborea*), Cheyanna Bone (*Honey Bee Allies*), and Elena Kfoury.
2   The lemniscate (∞), commonly called the 'infinity symbol', is an essential recurrent pattern of bee teachings (see Chapter 14).

Immortal Being

The bee colony is a superorganism composed of numerous organs and tens of thousands of bee-cells. An isolated bee does not have her own individuality and her survival is totally dependent upon that of her colony. The superorganism has, uniquely, the feature that its members are all interdependent while physically separated from each other. This characteristic gains itself an incredible malleability and a powerful resilience. The superorganism is an elastic being who can expand and contract itself at will, and is thereby capable of adapting to all sorts of the most complex situations.

Furthermore, a fascinating thing, the superorganism is *potentially immortal*. In effect, each of her members, her organs, her cells, is constantly and cyclically replaced. The average duration of the life of a queen is from one to four years, while that of males is from fifteen to fifty days. During the summer, the lifespan of maidens is similar to that of males, but during the winter, maidens' lives may stretch up to four months long.[3] Bee reproduction occurs at two levels: sexual reproduction within the colony and asexual reproduction at the level of the superorganism herself. In the bosom of the hive, the queen-mother lays the eggs of maidens, drones, and future queens, thereby constantly renewing her population. With respect to the colony herself, she reproduces by division, or swarming, which ensures the renewal of her genetic heritage. Therefore, as long as she does not encounter an accident on the way, the hive superorganism is maintained, permanently, in an eternal youth. This makes one dream, *n'est-ce pas...?*

The Voice of the Bee

"Each colony is a physical manifestation of her *Hive Being*. Each Hive Being bears her own history and has forged a personality during the course of her multiple incarnations. Some are older than others. Colonies with an older Hive Being are recognisable by the depth of their aura and the maturity of their exchanges with human beings.

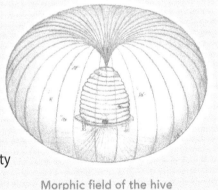

Morphic field of the hive

---

3   The commonly-adopted name of 'workers' fails to honour them; we will name our graceful young girls as 'maidens'.

The Hive-Being is a *morphic field*[4] which interconnects all the members of a colony. The Hive-Being's morphic field is similar to a web of luminous filaments which broadly encompasses the physical body of the colony. Even if it is invisible for most humans, it is clearly perceptible for the bees at every moment.

This *intranet* guarantees the cohesion and organisation of the colony. A multitude of information circulates by means of the threads of this canvas, which enables each member to be constantly up-to-date on the general state of their colony, its activities, and its needs... The proper functioning of this internal network is the foundation of the collective intelligence that unites us and guides us. The adaptability and the resilience of the colony is proportional to the state of health and of vitality of her morphic field."

The Organs of the Hive

> You are not a drop in the ocean
> You are the entire ocean in a drop.
>
> Rûmî (1207-1273)

Each bee has the capacity to receive, analyse, and interpret the information she perceives. However, the hive mind – her Being – is much more than a collection of tens of thousands of autonomous mini-minds placed side-by-side. *The Whole is greater than the sum of its parts,* the Bee Deva tells us. *It is the synergetic communication between individual intelligence and collective intelligence which creates the splendour of a hive.*

---

4 The concept of morphic fields (or morphogenetic fields; from the Greek *morphos* - form) appeared in the 1920s. More recently, it has been particularly championed by Rupert Sheldrake, an English biologist and author. The morphic field is a non-material generator of material form. Its influence extends into space and endures over time. It is a matrix of information which archives the memory of each being or form and guides their development. Morphogenetic fields are the specific morphic fields of biological organisms. Emotional fields and mental fields are also morphic fields, known as thought-forms.

The hive superorganism is made up of multiple components which the Deva calls *organs*. Some are animal – the maidens, the males, the brood, and the queen – and others are mineral and vegetal – the wax, propolis, honey, pollen, venom, and the hive body. Interconnected with the others, each element has its own role and its own history. In the same way that the internal organisation of the human body fascinates through its complexity and the magic that takes place in the coordination of its composites, the hive organism is incredibly coherent, ensured by a highly sophisticated internal communication system.

The Voice of the Bee

"The Hive Spirit governs a concerted ensemble of *Sub-Devas*. These are her organs, physical or non-physical: the Deva of the Queen, the Deva of Wax, the Perfume Deva, the Deva of Song... Each Sub-Deva has her own qualities and her *medicine*,[5] which she places at the disposal of Nature and Humanity.

Certain organs are polarised. Their spatial and temporal distribution are of major importance, on several levels. Honey, for instance, is negatively polarised (-). It forms a thermoregulatory insulating layer which facilitates maintenance of the internal temperatures necessary for brood development and the survival of the cluster in winter.

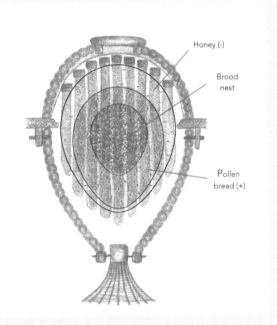

Honey (-)

Brood nest

Pollen bread (+)

Internal structure of the hive

---

5   The term 'medicine' relates to the definition employed by some Indigenous cultures of North America. The medicine of a plant or a substance is not limited to its phytochemical attributes; it also encompasses its energetic properties and its *teachings*, which is to say that which humans may learn from it and understand about themselves and the world.

It acts equally as a regulator of cosmic radiations (+) and filters energetic disturbances, such as the electromagnetic waves emitted by human technologies. Pollen bread, meanwhile, has a positive polarity (+). It protects and nourishes the brood, and regulates the flow of telluric radiations (-). This alternation of positive and negative polarities reinforces the stability and the resilience of the hive system.

Delicate coordination between the various organs of the system is governed by the quality of information circulation within the *feedback loops*. These loops interconnect the local and the global; information circulates from the morphic field – the Hive Being – towards the physical organs and inversely. Each organ has its own media for relaying information: the hexagonal network of honeycomb propagates vibratory information, the internal atmospheric network diffuses olfactory information, and the network of antennae disseminates pheromonal information... Thanks to the feedback loops, the Hive Being evolves over the course of her multiple incarnations; her cumulative memory allows for the expansion of her consciousness and the maturation of her personality.

The dynamic of the feedback loops is represented by the symbol of the lemniscate, wherein this flow of information cascades through the different physical and energetic bodies of the system. The effect of each action constantly reverberates through the global system, which in turn informs each organ of the consequences of its action, inviting it either to accentuate or to curb its behaviour. Through retroaction, the organ readjusts its behaviour in the common interest.

This manner of functioning is applied to all systems, whether they are macroscopic (the solar system) or microscopic (the atomic system). The efficiency of the feedback loops depends on the conductivity of the paths used for circulation. At the physical level, the greater the crystallinity and super-conductivity of the materials, the more any loss or distortion is minimised.

Here is an example of a feedback loop in a hive: let us imagine that we are in springtime, at the height of the lavender bloom, an abundant source of nectar. On this sunny morning, the number of foragers at work are 21,357, which represents 43% of a total population of 49,669 bees. At the heart of the hive, everyone *knows* it. This knowledge is not mental or theoretical, it is an *intuitive knowing*.

Now let us imagine that a great wave of brood are on the point of hatching. The nurses inform the Hive-Being that the need for nectar has increased. The collective intelligence of the hive will immediately adjust and new foragers will be recruited. If, at the same time, the handlers realise that the storage space for fresh nectar is insufficient, they send the feedback through the network and builders are engaged to build new comb. The feedback loops are all the more effective when the sources of labour are plastic: a supply of temporarily inactive bees is permanently available to be mobilised according to the various needs of the whole. While they are idle, the bees rest and refresh themselves in a meditative state.

Now let us imagine that a danger is suddenly detected in one of the sources of harvested nectar: the emanation of a toxic product applied by some farmer, for example. In this case, a 'stop' signal will be communicated through the network and the 'foragers' organ will incorporate it by modifying their behaviour and seeking out other, healthier sources of nectar.

The homeostasis of an organism is its capacity to maintain inner balance through constant adaptation and adjustment to its environment. It is assured by the fluidity and bi-directionality of the feedback loops. If this stream of information is hampered, distorted, or interrupted, the system enters into disequilibrium and declines."

Nature and Cosmometry[6]

Since I have been communicating with the bees in a regular and conscious manner, my vision of the world and of Nature has changed profoundly. The Bee Deva accompanies me during my experience of life like a firefly, an inner guidance which continually invites me to develop my *subtle perception*.

When I walk in nature, I am constantly arrested by the underlying cosmometry which governs the ordering of our reality. It is as though I am perpetually 're-minded' that all form,

6   Cosmometry is the totality of fundamental patterns, structures and principles inherent to manifestation at all scales within the cosmos, along with the science which measures and studies them. I prefer this term to 'sacred geometry', since, ultimately, isn't all geometry sacred? In addition, the scope of cosmometry is greater than that of geometry, which, etymologically, concerns terrestrial measurements (*geo* denotes the earth, in Greek).

whether spatial or temporal, is the ephemeral manifestation of a composition of sacred proportions. The trajectory of a bird on the wing, the manner in which a wave licks the boulder, the symmetry of flower petals, the whorls of mist on an early morning, the interferences of circles made by drops of rain on the surface of a lake... It becomes evident that all, absolutely all, unrolls from an omnipresent and universal mathematical mosaic. This perception of geometric patterns which encode the world, such as the harmonious ratio of the Golden Number,[7] calls me to reconnect with the essence of the world.

We perceive the physical dimension of the world through our five senses. The Bee Deva invites us to widen our spectrum of experience by *transcending* our senses; that is, to rise from sensory perception to extrasensory perception of the world. Clairvoyance, clairaudience and clairsentience are not gifts reserved for a select few. Everybody is able to develop this subtle perception in everyday life. Newborn babies and young children do it naturally, as they *feel* vibratory fields before perceiving forms. This intuitive receptivity, often unlearned while growing up, may be reactivated consciously and trained. Just like our physical muscles develop through daily exercise, it is possible to develop the muscles of our perceptive abilities. To *feel* the mood of someone without seeing or hearing them, to *know* that somebody is staring at our back, or to *sense* that the telephone is about to ring are many examples of this extrasensory sensitivity.

Beyond the basic equipment of our five senses, we have at our disposal a variety of internal antennae which are available for greater or lesser refinement, depending upon our affinities. Visually-oriented people will translate information more in terms of images and patterns, without necessarily using their physical eyes. Kinaesthetics will be more sensitive to information as it moves and resonates within their bodies, while auditory-oriented ones will transcribe it into sound *patterns*. The more we get into the habit of activating our antennae, the more the morphic fields and their interactions reveal themselves to our consciousness and the more our intimacy with the world deepens.

Girls of the Golden Number
Strong in heavenly law
Fall upon us and dream
A honey-coloured god.[8]

Paul Valery, *Cantique des colonnes*, 1922

The Golden Number (or *Phi* - φ) is a recurring ratio in nature, with a value approximately equal to 1.618.

8   Filles des nombres d'or / Fortes des lois du ciel / Sur nous tombe et s'endort / Un dieu couleur de miel.

The Voice of the Bee

"When you understand that the world of physical form is only a tiny part of reality, the true essence of the world becomes accessible to you. Despite their small size, bees perceive the world from a much larger point of view than the majority of human beings. *What you see depends on how you look.* When you look at things too closely, you only see a part of the whole picture; it is only when you back up that you can see the role played by each element of the system and how it is linked to the others. Thus, an increasingly complete picture of reality is revealed.

Our world is cosmometric; it is built according to an extremely precise and coordinated mathematic. Human vocabulary is still too primitive to describe this mechanism accurately. Every form, animate or inanimate, material or immaterial, is the result of a concentration of energy. Morphic fields are the underlying templates for the sequence of forms; they are fields of probability, which is to say that they are implicitly constituted by infinite possibilities which may be made explicit in the material world with greater or lesser probability. Every object, physical or non-physical, is potentially wave and particle at the same time. The more its *particle function* is active, the more its form is dense, localised and individualised. The more its *wave function* is active, the more it is universal and light, freeing itself from constraints of space and time.

Morphic field of the Bee Deva

Our Universe is a *holographic* system, composed of a multitude of morphic units nested within each other. Each unit, also called a *holon*, is simultaneously whole and part; it is the *whole* of a smaller system and a *part* of a wider system. Each holon contains all the information of the Great Whole. The dimensional ladder of holons ranges from subatomic particles to the multiverse (the totality of universes).

Each bee is a morphic unit of her Hive Being, which is a morphic unit of the Bee Deva, which is herself a morphic unit of the ensemble of Devas of the animal kingdom, and so on..."

3

THE SWARM: A WINGED SERPENT

Ethereal swarm, serpent on wings,
Intelligence incarnate, vast and free,
Treasure of life, expanding glee.

A Lesson in Humility

Thursday, 20 March

It is the first sunny day after a long period of rain, and a perfect day for my spring visit. I pause to say hello to each of the hives under my care, scattered around an area of several square kilometres. I weigh them one by one, assessing their strength, feeling their energy and the vitality they emanate, and I remove the entrance reducers.[1]

My circuit is just about finished. I feel an almost palpable vibration when approaching the last hive, located on the terrace overlooking Tamera's South Lake. An unusually dense and exalted hum tells me that something special is underway. I seat myself alongside, curious, and delicately pull away the entrance reducer. With that small gesture... Whooosshh! A wave of bees gushes out of the hive like a fountain of plenty! The hive swarms before my dazzled gaze – what synchronicity! If I triggered their egress with my action, I couldn't say. Either way, they were undoubtedly already prepared to launch.

This is the first time that I have the honour of witnessing a swarm's departure from the belly of her mother. I lay on my back, fascinated by all the patterns the bees are forming in the air. And to say that one among them, just one, is the mother of all the others.... This cloud protects her like an enormous cocoon. In the heart of this jubilant throng, she is invisible, untouchable. Each bee seems to be perfectly conscious of her position and that of her neighbours, as if the speed of their movements was precisely regulated by a collective intelligence. Blinking, I lose my gaze in the flow of geometric forms in perfect proportion unfolding a mandala on the deep blue background of the sky. This mosaic calls to mind the heart of a sunflower, whose vibrancy makes

The swarm in a cluster

---

1   I use wooden entrance reducers during the winter to support internal temperature regulation, and to prevent mice and other undesirable visitors from accessing the hive.

me dizzy when I dive into it... Suddenly, as though each bee has heard a signal inaudible to my ear, the flying multitude shifts northward like one single body. The etheric ball glides through the air with a winged serpent's graceful, harmonious movement. The swarm doesn't delay in clumping onto the branch of a small oak some thirty metres from the mother hive.

After this ecstatic wave outside of time, my rational mind starts to churn: "Gosh, I forgot my gear for swarm capturing, should I go and get it? I will have to be quick if I don't want to lose them! Will they, on their own, choose one of the swarm-attractors which I have sprinkled around the bee yard...?"[2]

"*Shhhh... Be calm...*"

I almost spasm, I am so startled. I am only beginning with the hives and this is the first time that the Bee Deva speaks to me in such a clear, distinctive manner. I calm myself and breathe, listening, amazed.

"*A swarm is a fragile being and this moment is crucial for the destiny of this colony. Avoid disturbing her with your anxious thoughts. We offer you a gift to contemplate, to live, to feel. Place your thoughts aside and gently come closer.*"

Heart pounding, I take a few steps toward the pulsating cluster that is organising itself after landing on a low branch of the small oak. I admire it in silence. Then I feel the Bee Deva take my hand and softly guide it to caress the cluster, warm and moving. My hands tremble slightly, then calm down; I close my eyes. Warm waves travel through my chest and my belly. A wave of love invades me; tears flow.

"*Now that you have adjusted yourself to the energy of the swarm, you may speak with her. If you wish, you may indicate a destination.*"

"Really?"

I have trouble believing it. Is such a technical communication with the Hive Being possible? To discover the answer, I have only to make the attempt. I visualise one of the swarm-attractors and inwardly declare, "Dear swarm, thank you for your trust. I invite you to install yourselves in this new home I have prepared, if it suits you."

"*All right. Now, have trust and go!*"

"Uh... Yes, okay..."

---

2   At the beginning of every spring, I install numerous swarm-attractors (also called bait hives) in the vicinity. In order to attract the scouts, I brush the inner walls of the empty boxes with a little melted propolis and apply citronella cream to the entryway.

The next morning, my curiosity is so intense that I almost run to the apiary. "Yes! It worked!" The swarm is quietly installed in her new residence. Their new life begins, the scouts orient themselves, the queen-mother lays, the workers build, and the choir sings with ardour and joy. As for me, I savour the magic of life...

Individuality and Collectivity

The preceding painting is an allegorical representation of the winged serpent formed by the swarm in flight, the one which inspired me so profoundly on that day. The unified living wave moves herself with an incredible coherence and coordination, similar to a *murmuration* of starlings.[3] The level of collective awareness is at its peak. Each particle is entirely and unconditionally devoted to service of the whole.

The concept of 'swarm intelligence' or 'collective intelligence' is very much in fashion among various branches of science and economics these last few years. Inspired by observations on the efficiency displayed by the behaviour of social insects, this concept encourages individual responsibility within a collective organisation. There is no leader who decides and commands the others. Rather, decisions are taken collectively and guided by a global, intuitive coherency. Individuality and difference become valued when they interact in synergy for the common good.

Losses and Gains

I feel saddened when I realise that, for most beekeepers, swarming is considered something to be avoided at all costs, a synonym for loss of yield and for 'waste'. Indeed, hives which swarm do give much less honey or even none at all. To avoid it, many beekeepers systematically destroy the queen cells. Financial gains for the beekeeper involves the loss of one of the most wonderful moments of joy and revitalisation for the colony. Difficult, so difficult to reconcile yield and respect... Which compromises to adopt, what new methods of relating to introduce? A difficult dilemma to confront if we hope to rely financially on the products of the hive.

---

3   *Murmurations* are assemblies of thousands of starlings wheeling together in the air. If this inspires you, it is easy to find videos of murmurations to watch online. They're fascinating.

The Voice of the Bee

"The swarm is an annual *diastole*, a great puff of fresh air and regeneration. The colony in transit achieves her maximum expansion; she swells with vital energy that revitalises her etheric body and reinforces her immunity. The swarm is an ecstasy, a celebration, and a festival of light.

The days preceding the flight are perfumed with an ineffable excitement. Members of the convoy are chosen from among the most experienced. The moment of take-off is carefully chosen for its physical, climatic, and astrological attributes.[4] For the first swarm of the year, it is the current Queen-Mother who will fly. Placed on a diet for several days, she is slimmed down so she may fly again.

The Queen-Mother delights in these annual moves, as, other than during her nuptial flight, they are the only moments of her life where she sees the light of the sun. This solar bath is like a new coupling, where the stock of sperm she received during her mating flight is revitalised.

The process of dividing the colony summons a new Hive Being to take form. Intimately connected to the geographical surroundings, the Hive Being of origin remains in the mother hive; a new Hive Being incarnates in the swarm which flies out. The physical expansion of the flying swarm *opens the door* to coalescing the new Hive Being. Her presence is then *imprinted* on the matrix of her new body when the bees gather to cluster. This is a crucial phase and a particularly vulnerable one in the life of a colony being born. The practice of artificially dividing a colony makes this process of incarnating more difficult for the Hive Being, since the expansive phase of flight is absent, but it is achieved nonetheless.[5] The swarm adapts herself to the offered conditions as much as she can.

---

4   The absence of rain and wind are favourable criteria. As well, I have noticed that swarms prefer to take flight on 'fruit' or 'root' days of the lunar biodynamic calendar (see chapter 8, footnote 1).

5   Dividing, or 'splitting' a colony means taking some of the frames and placing them directly into a new hive box.

During the search for our new home, which may last from several hours to several days, our scouts are the exploratory and sensory organs of the swarm. They comb through all the cavities in the area which have the potential to receive us. During her appraisal, while the cluster waits patiently on a branch, the scout bee projects a holographic image of her colony into the potential house. Numerous elements are considered; the fundamental criteria of selection are a cavity neither too large nor too tight, a sunny entrance of the right size, and a micro-climate neither too dry nor too damp. As much as possible, they will chose a sub-soil undisturbed by geobiological perturbations and a favourable position on the telluric network.[6]

The scout takes samples of the materials which make up the interior of the cavity by rubbing her body against them, so that members of the *Council* can feel its atmosphere. Upon her return to the cluster, she dances the information about her discoveries, including her personal appraisal of the quality of the venue visited. Each potential venue is the subject of numerous comings-and-goings by other scouts, each one bringing her own evaluation of the spot. The decision-making Council, composed of a hundred(ish) experienced bees, is the relay station of the colony's collective mind. Thanks to the feedback loops, a *quorum* is achieved naturally after a certain time, and the final decision is attained transparently, without conflict or effort.[7] The swarm joyously takes off once more towards her new home, guided by the scent of the scouts."

6   The telluric networks are vibrational grids that surround the earth. There are many grids of different sizes, intertwining in a complex latticework. Even though my research in this area is still very embryonic and my hypotheses unconfirmed, it seems to me that a swarm cluster, while she is in transit, likes to settle on a negative node of the Hartmann network, the lines of which are 2 to 3 metres apart. The location of the nest will preferably be established on a positive or neutral node. It is amusing to observe that ants and cats also seem to be more fond of the negative nodes. Further, Yann Lipnick, in his book *Géobiologie, Enseignements et Révélations des Gardiens de la Terre* (currently available in French only), writes that the preferred location of the bees are found on the positive lines of the Peyré network, which are spaced 11 to 14 metres apart. This network, resonant with the sun and called the 'sacred network', is connected with the heart chakra and the vibratory signature of gold and copper. As much as possible, the colony also will avoid installing herself above an underground stream or a geological fault.

7   Thomas Seeley, an American apidologist, describes this process with great precision in his book *Honeybee Democracy*.

4
LOVELY MAIDENS

Beautiful maidens, *belles demoiselles*,
Imbuers of love, and eternal virgins,
Devoted connectors, serving the Whole.

Adaptability and Resilience

Tuesday, 5 June

My neighbour this morning: "By the way, Sandira, the other day I noticed that the hive you installed behind my house has overturned.

– Overturned?

– Yes, I didn't dare to look closer, I don't know much more about it…

– Oh no! The swarm is only a few weeks old, still so fragile…"

Inwardly, I grumble my irritation about the time he took in letting me know, but I don't say anything to him. Then I get myself over there as fast as I can and, in fact, the hive is head-down in front of its base. Her freshly-made combs are turned upwards. Some animal must have tipped it over, probably a boar. Worried, I approach.

I am quickly reassured; the little swarm emanates a serenity which touches my heart. Despite the light rain which fell during the night, she continues her young life, adapting herself to the situation without drama. The foragers are bringing in nectar and pollen, and I even see one dancing.[1] The colony shows neither signs of aggression nor of despair; she has come to terms with the conditions offered by the new situation and life continues…

This adaptability inspires me. I realise how much lighter my life could be if I acquired this capacity to simply 'welcome that which is'. Suddenly, I smile at the choler I felt a little earlier towards my well-intentioned neighbour. Without him, the hive might have remained with her feet in the air for several weeks before I noticed it and put things right.

I replace the hive in good and proper position, and secure her stability. Two or three bees circle me calmly, as though to thank me. Without lingering, though, they return to their occupations.

---

1   The dance of the forager indicates to her sisters the location of a food source (see chapter 14).

The Value of Service

> Whatever you do, do it fully,
> do it with the fullness of your heart.
>
> Sabine Lichtenfels, *Sources of Love and Peace*

The maidens tend to most of the practical tasks in the hive system. They demonstrate total devotion towards the superorganism and are ready at any moment to adapt their service to the needs of the tribe, borne by a natural and joyous unconditional generosity. They offer the best of themselves to the utter limit of their capacity without ever begrudging it... And if our symbolic cultural baggage generally depicts the 'workers' as zealous and driven to excess, you should know that they also honour the importance of contemplation, meditation, and repose, as we will see below.

A curious and fascinating fact: the metabolism of the maidens changes with their age and the different tasks to which they apply themselves during the course of their lives. After her birth, the 'classic curriculum' of a maiden evolves her from cleaner to nurse, then handler, builder, fanner, sometimes guardian, then forager, and, for some of them, scout. This trajectory is not necessarily fixed, and it may happen that some older bees resume tasks of their youth if needed. Certain roles are accompanied by the acquisition of specific metabolic modifications: nurses are endowed with cephalic glands which produce royal jelly,[2] while the builders develop eight wax glands within their abdomen. The maidens' course through these different tasks is a great school of life...

---

2   The glands which produce royal jelly are housed in the head of the bee. Secreted by orifices situated in the hollow of the mandibles, this super-protein mix serves to nourish the young brood and the queen.

The Voice of the Bee

"During the course of their incarnation, maidens evolve physically, spatially, and spiritually in concert. This initiatory journey is deployed from the interior to the exterior, from the darkness at the heart of the hive to the light of the sun.

Maidens are born in the intense dimness of the brood nest, the place closest to Source. At birth, the young maidens are fragile. Slowly, they become accustomed to the earth's atmosphere and energy while remaining protected within the soft reaches of the nursery. They begin their adult lives with cleaning tasks, then they are initiated into taking care of their younger sisters who are still gestating in the cells. During this phase, they learn *to nourish the body of their Hive Being*.

Around the tenth day of their life, the maidens move towards the nest's peripheral areas, from which they perceive the first rays of sunlight which glimmer across the threshold. The task of building honeycomb stimulates their structural perception of the world. They learn to read the architectural plans of the matrix and to give them form in the material world. Contact with the Wax Deva infuses them with a vast amount of information about the activity and the history of the colony. They learn *to build the body of their Hive Being* by building the mortar of her skeleton.

The next job which summons the maidens is that of handler. When receiving the nectar, pollen, and water brought back by the foragers, the maidens come into contact with the environment outside of the hive. They learn to differentiate the various flower essences, they acquire knowledge about their properties, vitamins, and minerals. This knowledge will be deeply useful to them when they become foragers. But above all, they learn *to alchemise the body of their Hive Being*. Just like apothecaries in a laboratory, they discover secrets about the *sublimation* of gross material into subtle matter, how to transform lead into gold.

The following task is that of warmer or fanner. Engaging in this function is essential on several levels. On the physical level, the maidens develop muscles on their thorax and strengthen their flying muscles, an indispensable prerequisite for their coming mastery of the art of flight. On the spiritual level, as and when they improve their understanding of the art of maintaining thermal and respiratory equilibrium in the hive, they are initiated into *the sacred science of cycles* - fundamental knowledge which will then serve their shamanic work outside of the hive.

According to the needs of the colony, some maidens will become guardians. This forges their character, as they learn *to recognise the self from the non-self*, to say *'yes'* or *'no'* in a clear and authentic manner.

The apogee of their terrestrial development is manifest when the maidens become foragers, and they offer themselves with delight to the solar star. Now mature and skilful, they are able to physically distance themselves from the nest with ease, totally devoted to serving the colony and the *Gaian system* in which they know themselves to be an integral and co-responsible part."

Meditation and Quivering

It is common to gauge the vitality of a hive according to its eagerness to work. 'Activity = prosperity' is a well anchored cliché, a mirror of our society. However, it is not always what the Bee Deva suggests…

It happens that some hives in good health, with well-filled storehouses, will display a very strange behaviour: young bees, sometimes in great number, spread themselves around the entrance and on the inner walls of the hive, and rock, back and forth, while making a rotating movement with their front legs. This behaviour is called 'washboarding', after the traditional movement of rubbing laundry on a washboard.

With every animal behaviour assumed to have a practical function by the materialist researcher, diverse hypotheses have been put forward regarding its usefulness: cleaning and prevention against the development of moulds, the diffusion of pheromones to signpost the entrance, the recuperation of volatile esters deposited by the foragers upon landing… These may be partly true, but the Bee Deva here contributes clarification with a broader scope.

The Voice of the Bee

"Your science has now largely demonstrated the benefits of meditation on your brain and your health. It is described as 'promoting well-being, reducing stress and augmenting the capacity to concentrate'. It is the same for all living beings: every being meditates in different ways, even if only unconsciously during sleep. To meditate means to slow or to cease activity in favour of nourishing one's vibratory field, and to be nourished by it in turn. It is about creating punctuation spaces where the connection to Source is reaffirmed, diffusing a quivering luminosity within and around an organism. Source energy bursts vigorously from the heart of each cell, flooding the surroundings of the meditators.

The more a being has the opportunity to meditate, the more they thrive – but the role of these practices goes far beyond any benefits the individual may receive. Meditation is, above all, collective prayer, a manner of participating in the prosperity of All. When the maidens enter into collective prayer, a quivering comparable to champagne bubbles runs through the hive and spreads through the veins of Gaia. In proximity to a hive in meditation, the human being who is open to these frequencies may feel serenity and an infectious wholeness.

We practise many forms of meditation, more or less internalised. Here are a few examples:

- meditation in *isolation cells*: the maidens regularly withdraw from the collective hubbub by burying themselves head-first in a cell. There, in the most complete darkness and in contact with their Hive Being, they recharge their batteries, and acquire knowledge.

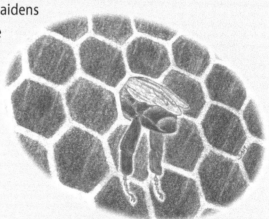

A maiden meditating in a cell

- *undulation*: the maidens gather on the inner or outer walls of the hive and enter into a collective rocking motion. These oscillations are induced by the cosmometric dynamics of the ethers, like the arabesques of the wind, the swaying of branches, and the coming-and-going motion of waves. Undulation links up polarities, which has the effect of regulating and synchronising the physiological and energetic functions of the system. Like the chirping of a cricket, the kneading of a cat, or a child on a swing, the bees' undulation revitalises the matrix and stimulates the circulation of our planet's blood.

- *dancing flight*: during fair weather, males and maidens of all ages like to fly vigorously in harmonious loops around the hive, sometimes several times per day. During these flights, the young bees test their inner compasses against the electromagnetic map of the area.[3] An intense joy accompanies this collective dance, which radiates widely around the hive. This communal exaltation is similar to that which humans feel when they gather for a collective event, a protest, or a celebration.

- *hygiene and touching*: like cats, chimpanzees, or dolphins, bees like to pass time rubbing, licking, washing, and touching themselves or each other. This is also a form of meditation, one which fortifies the integrity of the hive system and which cleanses the organism as much physically as energetically.

In a natural environment, when they can, colonies prefer to choose to settle in cavities of smaller sizes.[4] This way, when the honeycomb is full, they are more able to dedicate themselves to the intangible part of their terrestrial functions.

---

3   Those flights are commonly called 'orientation flights'. They may also serve as a cloud of camouflage protecting the princess in taking off and returning from her nuptial flight.

4   Thomas D. Seeley, in his book *The Lives of Bees* (see bibliography), demonstrates that bee colonies living wild in the Arnot Forest (New York state, USA) prefer to choose cavities averaging in volume from 10 to 40 litres. As for the conventional square hive box, it has an average volume of 40 to 55 litres.

If you observe a domestic colony *undulating* at the entrance of their hive and you add a super[5] to it, you will see that they immediately reduce or end their practice and go back to work. Like humans, bees prioritise their needs, and priority is given to the satisfaction of physiological needs and security. This is to say, the security of having a safe shelter with sufficient food and warmth to survive. At the second level of this ladder are social and intellectual needs (sharing touch, love, and information). The third level concerns spiritual and creative needs. The more the organism has the space to satisfy needs at the second and third levels, the more they are connected with something larger than themselves and the more they place their life in service to the Whole. Colonies who meditate are generally calmer and more sociable.

When an empty space is suddenly created by a beekeeper inside the hive, the colony comes down the ladder of needs again and focuses its energy on filling this emptiness. If the beekeeper is anchored in a system centred on the maximisation of production, the colony adapts as best she is able. However, this material focus will only be endurable if there are a sufficient number of other colonies nearby to carry out the non-physical parts of our work."

---

5   A super is an extra box added to the hive body, intended for honey harvest.

5

QUEEN-MOTHER

Our queen is our light, our beacon, our heart,
Sovereign incarnation of Great Mother Goddess,
Clear emanation of our presence on Earth.

A Painful – but Instructive - Learning Experience

When I was starting out with my first hives, my contacts with the Bee Deva were fragile and uncertain, and I frequently allowed myself to be influenced by the advice of other beekeepers. One day, one of them suggested that I 'mark my queens' in order to recognise them more easily and to protect them during work in the hive. The operation consists of marking the thorax of the queen with a dot of paint whose colour makes it possible to identify her age. I wanted to attempt the experiment... and here is the story.

Wednesday, 22 May

For my first attempt at marking, I have found a yellow *Poska* marker. Once the queen is spotted, I move feverishly to trap her, hoping not to injure her with my clumsy hands. Her energetic radiance is so strong that my hands are shaking terribly. I open the lid of the marker, apply it to her thorax, and then - *"Oh no!"* The marker, having heated up in my bag, delivers more ink than anticipated. It drips down between her feet. Biting my lips, I replace the motley queen in her nest...

Tuesday, 11 June

Three weeks have passed and it seems to me that I am perceiving a drop in energy in the colony. This morning I decide to open the hive once again, to see and to understand. It is impossible to find the queen I marked and, with sorrow, I notice a cessation of egg-laying... It is clear that after my awkward intervention, the colony decided to eliminate this bad-smelling, cosmetically disfigured queen and to replace her with a new recruit, likely newly born and still virgin.

Although the colony continued to enjoy a prosperous life after this event, I was profoundly affected and decided not to repeat the experiment. Putting myself in the queen's position, I felt how this practice - like so many others - lacks respect for the Bee being. This mechanisation of the living reminds me of labelling prisoners with numbers. I feel deeply saddened when I realise how much the relationship with Nature is disconnected from the most simple values of respect and recognition. Our agriculture is based upon exploitation, profit, 'taking' and not 'receiving'. It is time to right the balance...

Incarnation of the Mother Goddess

The Queen bee is both *Mother* and *Sovereign* of the colony.

She is Mother to the thousands of children, male and female, which compose her community. For most of her days, she lays eggs one by one into the hollow of each hexagonal honeycomb cell. During periods of expansive growth for the colony, she can lay up to two thousand eggs per day, equivalent to her own weight! The Queen-Mother is able to choose the sex of her progeny: fertile eggs engender females and infertile eggs produce males. The collective intelligence of the hive determines the relative need for males and females. The builders construct cells of different sizes according to their destiny. Depending on the needs of the colony, the Mother will be guided towards narrower cells for her daughters or larger ones for her sons.

This remarkable genitrix is also a Sovereign. Not in the sense of an authoritarian monarch dominating a people of submissive subjects, far from that, but in the nobler sense of the definition: she emanates a sovereign grace which unites her people into a coherent and harmonious entity. For the sensitive observer, a hive without a queen is easily recognisable. There emanates a kind of dull and haunting buzz which our emotional vocabulary could describe as a sad lament. The aura of the orphan hive is dramatically reduced or even non-existent. At the opening, I notice that the movements of the bees lack coordination: they are jerky and their behaviour is incoherent. Although they sometimes attack in a sudden and frenetic manner, they generally ignore my presence, having nothing to defend.

The Voice of the Bee

"The Queen-Mother is our beacon, the heart of our Being. We belong to her as she belongs to us. Should she disappear, our existence loses its meaning. The Queen-Mother is our *uterus* in the vast sense; she is the crucible of Life with a capital L. As an interface between Spirit and matter, she is the fertile substrata within which the spark of Life opens a path towards form.

The Queen-Mother is the repository of the Bee Deva's ancestral wisdom. Stockpiled on the shelves of her morphic and genetic library, she conserves the annals of the 'Bee Code'. That is to say, all the information relative to the history of the bee since her arrival on Earth and beyond, as well as the specific history of the lineage in which she has incarnated.

The princesses are raised with great care. The egg potentially destined to become a queen is laid in a large cell curved towards Gaia, inviting her to incarnate. The larva is nourished exclusively with royal jelly produced by the maidens. When the princess hatches, she slides out of her case and turns towards the sun. She becomes acclimatised to the atmosphere of the hive before launching forth towards the sky on her nuptial flight.

Once she is fertilised, the Queen-Mother emits a rhythmic pulse which spreads through the hexagonal network of honeycomb. This allows us to locate her spatially at every moment. She diffuses around herself a reassuring and cohesive perfume. We cover our bodies with this precious fragrance; it circulates from antennae to feet, from feet to antennae due to our constant touching. This pheromonal unguent is unique to each colony and serves as an aromatic 'digicode' which guarantees entry to the hive by the foragers when they return from their collecting. If the Queen-Mother should suddenly disappear, propagation of the compound is interrupted and all the members of the colony are informed of it within twenty minutes. If it is possible, a replacement procedure will be initiated as quickly as may be."

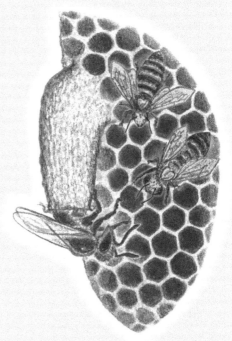

Hatching of a princess

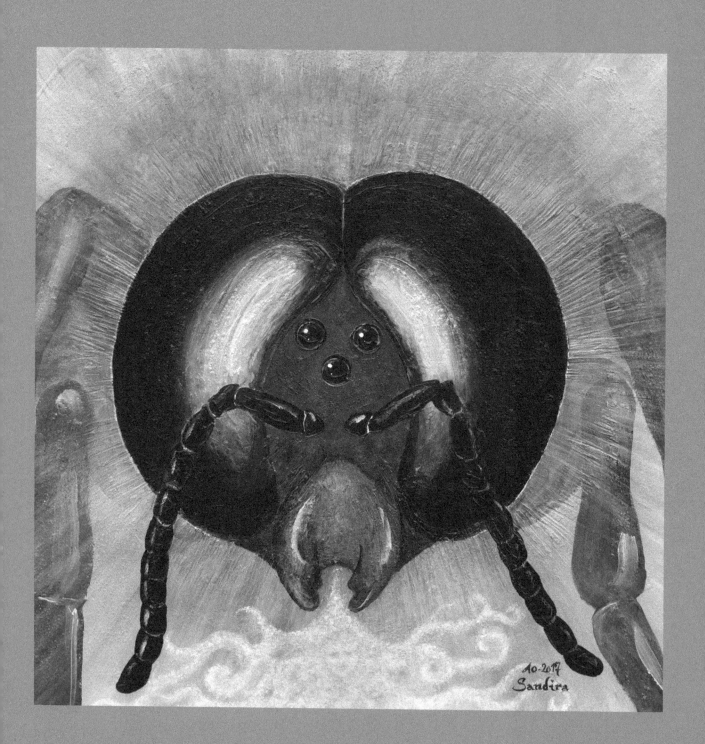

10-2017
Sandira

6

SONG OF THE MALES

Gatherers of secrets, history's trustees,
Senses of the hive, singing the sacred,
Innocent travellers, cosmic and free.

Sensitive Cells

Thursday, 24 May

A tickle awakens me and I smile.

I had dozed off, lulled by the soft humming of the hive in full activity. The tickler is a big easygoing male who has landed on my arm, seemingly curious to explore me. An erotic tingle sweeps through me. He feels his way across my skin with his delicate antennae. His two enormous eyes almost cover his entire head. He comes to a standstill for a moment, quivering, organs of perception on alert... Is he communicating with the Hive-Being? This is not the moment for intellectual questions, I let my eyelids droop closed again under his caresses.

Restoring Their Honour

In French, the male bee is referred to as 'faux-bourdon' (false bumblebee) - such a wretched term... On top of that, he has a terribly bad reputation. He is usually described as a lazy profiteer and a glutton of honey whose office has been reduced merely to fertilising the queen. Moreover, due to his 'mania' of visiting nearby hives, he has been accused of making a significant contribution to the spread of varroosis.[1] The Bee Deva views things otherwise: males are indispensable and sacred, as much as the queen-mother and the maidens.

Drones in flight

---

1   The illness of which the vector is the varroa mite (see Chapter 20).

The Voice of the Bee

"In tandem with the Queen-Mother, our males are the mobile repositories of our history. The knowledge that they hold speaks of the past, of the future, and of inter-communitarian relationships between colonies. They sing the song of the ancestors, as well as the history which we share with humans. They are responsible for circulating this knowledge throughout the nest, and for transmitting it to the brood with their songs.

Our males are the most sensitive sensory organ of the colony. Thanks to their big multi-faceted eyes, their delicate antennae, their extra-sensitive fur, and their intuitive intelligence, they behave like nerve cells, gathering and relaying information from within and around the nest that is essential for our evolution. The quality of their work shares in the social homeostasis which is established within the tribe. While the maidens occupy themselves fruitfully with the physical environment of the hive, the males explore other energetic realms of the terrestrial system. Whatever the male senses during his peregrinations, all the other inhabitants of the hive sense it also, instantly. The power of the drone resides in his inner silence.

They possess neither sting nor venom. They are completely inoffensive and know neither private property nor affiliation. They have nothing to defend, nothing to prove; they simply emanate the joy of being alive, welcoming that which is. When a male knocks at the door of a neighbouring colony, he is welcomed with open arms. This is not the case for maidens, who are refused at the entrance as long as they are permeated with the pheromones of their own Queen-Mother. Like foreign affairs envoys, the males weave social bonds with neighbours, inquire about the latest news, and transfer this information to their own Hive Being. They act as mediators, and their cohesive inter-community role is fundamental."

Droning the World

The English word for the male bee is 'drone', from which derives the name for these small unmanned aerial vehicles. Very useful for mapping the land, these devices capture splendid images of our beautiful planet and take high-precision measurements. In a similar manner, males offer to the hive an enlarged and extra-sensitive gaze at the world.

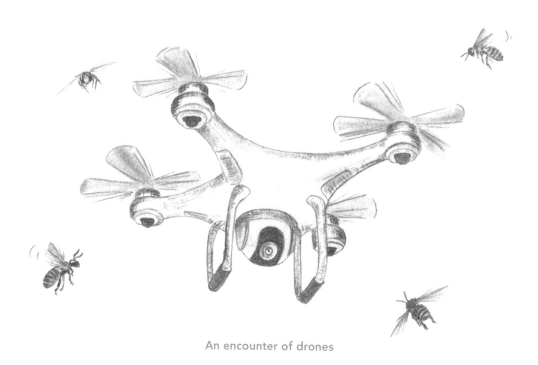

An encounter of drones

An aerial cartography enthusiast, my friend Joshua sends his drone one day to quarter the sky above 9 hectares of land recently acquired by friends on the coast of the Portuguese Alentejo. The data received by the apparatus will help them to understand the ecological dynamics of the land, among other things, and to set up a regeneration plan that will heal the wounds of monoculture and honour the profound potency of the place.

Suddenly, even though it is making headway through open skies, the device makes a series of unexpected stalls and the remote control repeatedly displays the message 'obstacle'. Joshua manages to land the engine and, finally, to understand the situation. "The drone was surrounded by a swarm of about 100 bees!" he tells me with a laugh. I laugh with him, but remain sceptical about the fact that it really is a swarm. In fact, a swarm is more heavily populated and moves clearly towards a pre-determined destination while it is in flight, without dawdling *en route*. When Joshua sends the video of the landing to me, I slow down the image and enlarge it, and note with pleasure that my intuition was correct: it is a group of drones that swirls around the engine, apparently deeply interested by this curious apparatus. By chance, Joshua took up a position at the level of the males' congregation zone, something that is not so common. In these places of particular energy, which the Deva calls *lumens*, males from nearby colonies gather together and impatiently await the arrival of princesses in order to live the ecstatic experience of coupling in flight – of which we will see more in the next chapter.

Temporary Disembodiment

Towards the end of summer, earlier or later in the season depending on the weather and the year, we observe a curious activity at the entrance of hives: most of the males are being driven off by the guardians and die from hunger or cold. It is not unusual to see a carpet of corpses covering the soil in front of the hives. Happily, insectivore and necrophagous allies (wasps, ants, birds, and spiders) come and feast on this abundant food. A small percentage of males will still be maintained during the autumn months, in case of a necessary fertilisation for the occasional replacement of a queen.

The Voice of the Bee

"When the expansive phase of the year draws to a close, the usefulness of males in their current physical form diminishes until it completely disappears in winter. The spirit of the males withdraws into the un-manifested dimensions and their physical envelopes are eliminated. Under the emotional eye of a human being, this procedure may seem cruel. However, it is nothing other than a natural adjustment adapting to the physical needs of the organism by avoiding an unnecessary loss of energy. Dear humans, your body also permanently eliminates thousands of cells which it no longer needs!

The presence and the song of the males nonetheless continue to resonate at the heart of the hive from the subtler spheres, awaiting their return to the physical world the following spring."

7

MATING FLIGHT

Princess ascends to the cloud of males,
The bridal flight throws open a door,
Heaven and Earth ecstatically create.

Princess and Queen

Thursday, 24 May

This morning, we are transferring a young, three-week-old colony to my neighbour's *top bar* hive; we have named her *Patience*. Given her size, I thought she was a primary swarm but observation disproves my assumption; while scrutinising the frames one by one during the transfer, I see neither eggs nor brood. Mmmm... Is this an orphan colony? The answer does not take long in coming, for then I see her: the *princess*, innocent and light.[1]

The princess is not a queen and the difference is notable. Physically, she is thinner; only the length of her abdomen distinguishes her from a maiden. However, the most remarkable difference lays in the energy that each of them emanates. When I have the opportunity to observe and *feel* a queen mother, it is much more than a simple insect body before me. The queen mother vibrates with an extraordinary power. Her aura illuminates the darkness of the hive like a full moon at night. She shines with a sovereign grace which inspires respect and admiration. She is always surrounded by her retinue: a crown of maidens who nourish and protect her at every moment.

The princess of *Patience* has neither court nor crown; she is as innocent and fresh as the morning dew. Worried that our princess is slow to mate, I call my friend Harald Hafner, a long-time biodynamic beekeeper, to ask his advice. Could the princess be infertile? "Patience," he says. "When the weather is rainy during the mating season (which had been the case for several weeks), the princess of a new swarm may not be fertilised as quickly."

Patience... Well, this hive is aptly named!

---

1   The primary swarm takes off with the old queen of the mother hive. If the queen is a princess, it is an afterswarm (a second or third swarm, or more).

The Voice of the Bee

"When the princess hatches, she is incomplete. The completion of her Being as Queen-Mother will only occur when she is fertilised during her nuptial flight. The mating flight is fulfilled in special locations called *lumens*.

The Sun and the Earth both emit light: that of the Sun is white and visible, while that of the Earth is *dark* and invisible to the superficial gaze. The lumens are situated at several hundred metres above specific points of the telluric network, where a concentrated flow of dark light erupts up from the Earth and fuses with the white light of the Sun. These places are acupuncture points situated along the meridians of the Earth. They shine with power and are perceptible to the young princesses and the males from several kilometres away.

When the sun is shining, our delicate princess draws close to the hive's entrance. She takes the time to accustom herself to the intense light of day, and then rises, magnetised by the nearest lumen. A cloud of maidens accompanies her take-off, to encourage her and to protect her during the first few metres of her journey. When she penetrates into the lumen, where gravity is reduced, the princess, now extremely light, is transported towards the Cosmos. Like a bridal veil, she diffuses behind herself a trail of intoxicating erotic scents which draw up the males from the surrounding colonies in pursuit of her.

The most agile and rapid males will have the advantage in this race. When penetrating the sacred antrum of the princess, the male inserts his phallus like an offering. His empty envelope then detaches itself and falls to the ground like a faded petal. All his vital force enters into the princess to crown her Queen; he dies to create Life. The corpse which detaches is no more than an empty vehicle which has achieved its function, in the same way in which the tail of a spermatozoon remains outside the exterior of an ovum during the fertilisation of an egg in the uterus of a woman.

Fertilisation

If the day is fecund and the princess lucky, twelve to twenty-four males will deposit their semen in the royal temple. Each new embrace is an ecstasy that radiates sparks like fireworks.

This erotic celestial encounter goes far beyond a simple physical coupling. There are multiple repercussions at many levels. The lumens are energetic portals for receiving and transmitting information. The field of the lumen is impregnated by the energy generated during the act of union and transmits this energy into the interconnected telluric and cosmic networks. The ethers of the Earth and the Cosmos are invigorated and regenerated. Similar effects are produced during sacred sexual intercourse between human beings.

The Queen-Mother, for her part, is now complete and ready to assume her maternal and royal offices at the heart of her colony. She will conserve her reserve of semen for many years; it will be revitalised each time the queen sees the sun again when swarming."

Tuesday, 30 April

I ready myself to open the hive. For a moment, I observe the flow of foragers who tranquilly come and go, their wings bright with the rays of a setting sun. Little fairies of light, they string behind themselves a trail of golden powder.

On the temple threshold, a small ball of bees suddenly appears, tightly compressed one against the other. The ball is so compact that I can make it roll by pushing it with a finger. I suspect that something undesirable is inside it, something in process of suffocating.

A few minutes later, the ball falls to the ground and breaks up. As the bees calmly return to the nest, I discern, immobile on her back, a princess laying dead on the earth. Two maidens meticulously lick her envelope, gathering, it seems to me, the last traces of her vital energy.

Curious, I delicately open the hive. It is a very small colony who has only just built three little combs. Going over the frames one-by-one, it doesn't take me long to locate the queen-mother with her radiant aura. Assuredly, she has just returned from her nuptial flight. Overflowing with fresh semen, she prepares to lay her first eggs.

I ask the Bee Deva to explain the situation to me. *Secondary swarms frequently fly out with several virgin princesses in order to augment the chances of fertilisation*, she tells me. *However, there is only one spot for the role of laying queen; cohesion of the community depends on it. A human woman would not know how to function with two uteri, would she? It is the same for the colony. And so, the coronation of a princess involves putting an end to the others.*

The fascinating ability of the hive to regulate herself enchants me. From all evidence, it is the Hive Being, the control tower of the colony, who decides, organises, and supervises the regulating process of the reproductive system of the colony.

The elimination of a spare princess may take place through suffocation, as in the above example, or by stinging. Princesses and queens possess a smooth dart, different from the barbed harpoon of the maidens. This allows them to inject their venom without detaching their stinger, thusly eliminating another princess or queen without losing her life.

In the online Larousse encyclopedia's article on bees, I read the following extract: "As soon as she is born, the first queen rushes to her rivals to sting them to death. If several queens are born at the same time, a fight ensues until the best one wins, the vanquished being doomed to death."

While the tendency is in decline, I notice numerous authors continue to attribute behaviours and adjectives to the bees that are very emotionally charged. Nevertheless, the regeneration process of the colony, much more vast and complex than that, involves no hostility. The Bee Deva's discourse is denuded of all anthropomorphism. Beyond notions of good and evil, the logic of the living does not know *drama*. That which lives, *is*. That is all.

The Voice of the Bee

"A colony cannot survive more than a few weeks without a fertile Mother. Beyond that, her coherency falls apart. As well, the creation and crowning of a new Queen-Mother is a complex and wisely-orchestrated procedure. It is also a moment of celebration, since it entails a pleasant revivification of the erotic body of the colony.

The Hive Being impels the creation of a new Queen-Mother in four types of situations:

- when the current Mother grows old, is wounded, or has a decline in fertility;

In this instance, many royal cells are constructed with great care, generally on the side of honeycomb. In each of them, the old Mother deposits an egg that she infuses with her royal blessing, with *intention* for it to give birth to a princess. If the old Mother naturally ceases to lay, she is usually protected. She finishes her life as a Grand-Mother, emanating an aura of wisdom around herself.

- upon the sudden death of a Mother;

If, by chance, the Mother laid some fresh female eggs before dying, the colony has the opportunity to create a replacement mother from an egg originally destined to engender a maiden. This 'emergency' queen-mother is usually weaker than a queen-mother who has received the royal blessing from her conception. Sometimes the replacement will be, in turn, rapidly replaced in order to bolster the solidity of the throne.

- in the Mother colony, when she casts her first swarm

When she has the opportunity, a colony swarms every year or every second year. The current Queen-Mother flies out with the first swarm. At the moment of swarming, several royal cells are usually ready to hatch, to assure a rapid replacement.

- for the swarms that follow the primary swarm, if there are

If the Hive Being opts to continue swarming, then many princesses will take off with each new birth. A communication system of auditory, olfactory, and vibrational signals guarantee that the process unfolds in a good way".

The Mysterious Song of the Queens

To have the pleasure of hearing queens sing is an uncommon and very special experience. I perfectly remember the first time I had this honour, when I was an adolescent. I was helping my mother to put some empty supers on her hives in the mountains of Arrée, in central Brittany. We rejoiced at sharing this sweet and warm spring day, in the living bath of bee energy. We were at the end of our trip, jovial and perspiring under our veils.

Suddenly a series of powerful cries burst out of one of the hives. We were both stunned, frozen, as though time had stood still for a moment. The queen's song seemed to us as though it emerged from a world between the worlds.

For a moment of eternity, *doing* gave way to *being*. Struck with enchantment, mother and daughter looked at each other and *saw* as never before. I felt the immense love my mother dedicated to the bees. With profound gratitude, I understood the privilege I was given to be born from her womb and to have been impregnated with respect for the natural world during my childhood. In the depths of my mother's eyes, I felt shining those of my grandmother, those of the mother of my grandmother, and so on, into infinity... Today, I perceive this moment as a rite of passage, during which the song of the queen was the symbol for the transmission of a torch of matrilineal wisdom.

The Voice of the Bee

"The song of the princesses occupies a particular place in the complex communication system which accompanies the establishment of a new queen. Her energetic role is just as important as her physiological role.

Princesses and queens practise their songs at different points in their evolution. When they begin to sing, the material world freezes for a moment. The emission of a queen's song rekindles the illumination of a network of luminous fibres which spreads over the surface of the planet. The maidens surrounding the princess or the queen who is singing suspend their activities for several seconds and an energetic portal opens. The royal song resonates with whalesong, with that of the wolves, the owls, and other guardians of Gaia's fertility.

At the physical level, the song of the princess or the queen transmits information about her state of growth, her genetic makeup, and her vitality. Princesses usually come from the seed of a variety of fathers; when they sing, their characteristics are expressed and circulate through the air, the forest, and the honeycomb.

Princess who are still in their cells emit a series of 'quack' following a sequence of specific codes. Those who have hatched express themselves with a sort of 'toots' at a different frequency and rhythm. The information shared through their songs permits, for example, the maidens' regulation of the princesses' gestation period. It may be, for example, that it becomes necessary to delay the birth of some princesses by a few hours, or even a few days, to time their hatching with the lift-off of the next swarm."

A Fragile Fertility

I have lived in the heart of Alentejo, an arid region in the south of Portugal, since 2011. At the time of writing (October 2018), I am taking care of some twenty hives distributed around various eco-projects in the vicinity. Since I began to care for the bees in 2012, I have noticed a growing phenomenon: more and more, natural swarms are orphans or are endowed with queens of failing fertility. Several people in France and Spain have mentioned similar observations.

Numerous factors are the cause. Male side, their brood is often intentionally destroyed by beekeepers[2] and their fertility is reduced from the impact of pesticides. Female side, the practices of queen-rearing and the artificial insemination of queens are accompanied by notable secondary effects. These days, more than 80% of managed hives around the world are endowed with artificially-raised queens who are replaced every year or every second year. Genetic traits are selected for productivity, docility, and a low swarming rate. But these queens are so fragile…

---

2   One of the techniques for reducing the pressure of the varroa mite is to eliminate male brood. This practice is especially used in organic agriculture as an alternative to chemical treatments.

Artificially-reared queens are 'emergency queens'. The procedure consists of stressing a colony by eliminating their existing queen and demanding they create replacement queens in bulk. These emergency queens are created by the bees from maiden eggs placed by the beekeeper in artificial royal cells. In a natural colony, the mechanism for creating an emergency queen is very practical and permits the colony to survive the sudden disappearance of the existing queen. The larvae of these ordinary eggs are intensely fed with royal jelly, thereby producing princesses. However, these eggs – which are not destined to become queens at the moment of laying – are weaker than those of queens consciously laid. These fragile princesses (virgins, fertilised or artificially inseminated) are then isolated in a cage with a few bees for company, and distributed by postal services to the four corners of the earth. It's not surprising that their candles burn only briefly.

In 1923, during one of his lectures about the bees, Rudolf Steiner said "Today, it goes without saying that in certain respects one can generally only sing the praises of artificial breeding. [...] but this joy will not last for a hundred years. [...] We are unable to establish between the commercially-purchased queen and her workers this profound affinity which is established when the queen is she whom nature provided. But at the very beginning, this is not seen."

Generation after generation, strains of indigenous bees have developed characteristics adapting themselves to the climatic, geographic, and energetic conditions of their local environment. While imported queens carry genes stimulating honey production in their descendants, at times they fail to offer any resilient qualities in the face of local diseases or temperatures of the region to which they are introduced.

However, the primary causes of these queens' fragility heavily emphasised by the Bee Deva are their disconnection from the Hive Being at birth and the absence of fertilisation in the open sky. Those processes are essential for them to fully and nobly exercise their role as Mother and Sovereign.

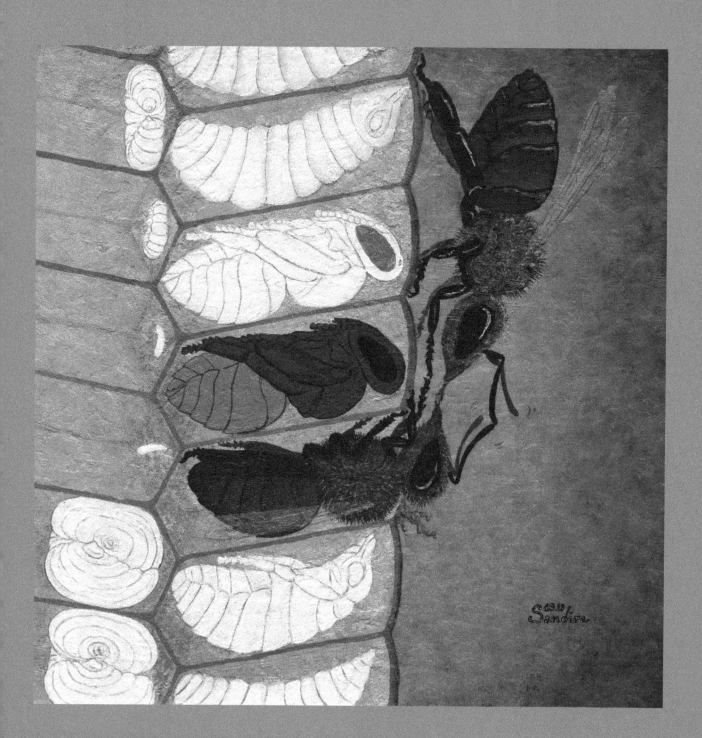

8

THE NURSERY:
PRECIOUS SANCTUARY

Egg into larva, larva to nymph,
Transforming I hum in my cocoon of love,
Into adult I mature, ensuring the future.

A Surgical Operation

Wednesday, 20 September

This last April, an enormous swarm installed herself in an old swarm attractor placed in the midst of some cistus shrubs. When the choice is available, a swarm prefers to choose an old weathered hive body over a hive made of new wood. The old box, broken up and blackened by preceding generations of tenants, had seduced her.

The colony exudes a powerful serenity; she is baptised *Kuan Yin* – goddess of compassion. With my friend and apprentice Annelieke, we would like to offer her a more watertight and solid home for the winter, and so we decide to transfer her into a nicely painted new hive. The day is chosen with care: sunny weather, without wind, and the moon in a 'root day'.[1]

We approach the colony with a soft tread. Five or six metres from the hive, we pause, ask permission to enter her aura, and express our intention to her. *Knocking on the door* is the least of courtesies when we visit friends, *n'est ce pas*? The reply is not long in coming: the energetic doorway opens and draws us subtly forward. A pleasant feeling of *welcome*.

With a mixture of dried local plants and a few pieces of propolis, I light up the smoker. Smoke with a balsamic aroma perfumes and consecrates the space like incense in an Indian temple. We *smudge*[2] each other to cleanse and settle our energetic bodies. The smoker is placed near the hive and releases graceful spiral coils of smoke throughout the operation.

We know that the process before us is delicate – even surgical – because it involves opening the brood nest. We have brought protective bee jackets with us if needed, but for now they are unnecessary as we feel ourselves to be warmly welcomed.

---

1   Solar, lunar, and planetary cycles have a notable influence on the climate, the energy of the Earth, and the mood of bees. I have been using the lunar calendar (or biodynamic calendar) as a daily planner for several years. There, I record my observations and explore the relationships between Bees, Earth, and Cosmos. According to the position of the moon and the planets in the sky, the influence of different ethers (earth, water, fire, and air) are more or less reinforced. Thus, a 'root day' indicates a particular affinity between the celestial bodies and the element of earth; these days are propitious for calm and grounding, physically as much as psychologically and spiritually.

2   'To smudge' is a practice which consists of filling the aura of a person with the smoke of medicinal or aromatic plants. This has a calming, cleansing, and protective effect. Only rarely do I use a smoker inside the hive; employed to excess directly on the bees, smoke has more of an oppressive effect than a calming one.

We put the hive in front of its supporting base and position the new box at the location of the old one. The frames are transferred one-by-one into their new space, gingerly. Moving the edge frames does not provoke a defensive reaction, but as we approach the centre and touch the frame with brood, tension escalates... The buzz coming from the cloud of bees around us becomes denser, although without being aggressive. *Take care*, they are saying to us, *this is our most valuable treasure!* Presence and precision are essential, as the slightest too-abrupt gesture could set off a colossal attack... When the last frame is shifted and the new home of *Kuan Yin* is closed up again, the Hive Being calms. The thread of trust has been preserved throughout the work. A feeling of profound mutual respect arises.

Once more, I notice how much my inner state is reflected in the behaviour with which the bees respond. Fear and chaotic thoughts will be reflected by a defensive attitude. Calm and trust engender trust and calm... which engender calm and trust... and so on…

In the Intimacy of the Brood Nest

The brood nest is, along with the queen-mother, the most valuable and sacred organ of the colony. Only very rarely do I open the nursery of my hives, conscious of the intense impact this has on the colony. It's one of the reasons why I appreciate the horizontal top bar hive, which allows one to open the side frames without affecting the brood nest.[3]

When I close my eyes near a hive and connect heart-to-heart with the brood nest, I feel infused with a warm and reassuring energy, soft and powerful at the same time, which calls forth a profound humility within me. I am transported beyond space and time, into the unfathomable waters of the lake of Creation. I am back in the womb of my mother. Like these thousands of little bees about to be born, I swim between worlds, nourished by the maternal placenta and rocked by the song of the Earth.

---

3   Contrary to those of conventional square hives (like the Langstroth hive), the upper bars of the horizontal *top bar* (also called the Kenyan hive) touch each other. Thus, when we open up the roof of the hive, the brood nest's inner atmosphere is not directly exposed to the open air.

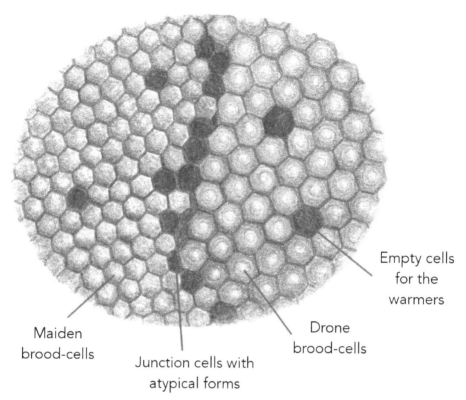

Empty cells
for the
warmers

Maiden
brood-cells

Junction cells with
atypical forms

Drone
brood-cells

Brood cells

The first few days after an egg hatches, the bee larva is nourished with royal jelly, this elixir of life which stimulates her immunity and connects her with her colony's Hive Being. Following this, the larva is fed with 'sororal milk' (made by their sisters), a mixture of pollen bread and honey which allows females to multiply their weight by 900 times and, for males, 2300 times over the span of a few days. After having accumulated physical mass, it is time for the subtle maturation which occurs during their metamorphosis into the state of nymph. The cradle-cell of wax is capped, which isolates them from the vibrant activity of the nursery until hatching.

The male brood are easily recognisable: their cells are larger and their caps bulge. In a healthy colony, 5 to 10 percent of the cells in the brood nest are intentionally left empty. The warmers slip into them and activate their wing muscles to spread heat into the nearby cells.[4]

---

4   Thermoregulation systems in the brood nest are described in detail in the book *The Buzz about Bees* by Jürgen Tautz (see bibliography).

The radical transformation of the larva – so simply constituted – into an adult individual – so complex and perfected – is fascinating... During this marginal phase of the world, the physical structure of the being in development is 'melted' and remodelled anew. The change of form underway is accompanied by a profound spiritual transformation of the being to be born.

The Voice of the Bee

"The nursery is our *Holy Grail*, the most precious sanctuary of our temple, from which flow the ties that bind us to Source. Calm, discipline, and hygiene are the order of the day. During the entire laying season, the nest is perpetually maintained around 34 or 35°C – a temperature close to that of your blood, dear humans. The air is very carefully supervised, the chorus orchestrated with deep awareness.

The opening of the brood nest by a human who ignores this careful orchestration is particularly intrusive. The impact of solar light directly upon the brood influences their development and their apprenticeship. It takes us many hours of work, sometimes even several days, to restore its atmosphere. Be aware that every opening of the hive's heart is a surgical operation which must be made with the greatest care: adequate preparation, contact with the Hive Being, rigorous hygiene, and inner calm.

The sacred space of the brood nest is a sanctum of sweetness, lulled by the interlaced songs of nurses and males. Their songs transmit the subtle teachings necessary for the initiation of the descendants coming to birth. The maidens sing of the present, the odour of flowers, the curve of the horizon and the temperature of the wind, as well as the essence of the tasks which earthly life holds for them. They infuse their songs with spatiotemporal and cosmo-telluric maps of their territory. As for the males, they are the storytellers: they sing of the past and of the future, the murmur of the ancestors and the contract which connects us to Humankind. The male and female hummings combine into a perfect symphony, a lullaby both heavenly and down to earth, accompanying our small ones during their process of incarnation.

Each baby bee who incarnates is unique, and carries within them the promise of tomorrow. The specific characteristics and abilities of the newcomers vary according to the genetic information received from their father and according to the care which they receive during their gestation. The diversity of male semen which penetrated the Mother guarantees the vitality and the resilience of our future progeny. Each maiden refines particular traits in her cell. If she develops an affinity with the ethers of water, she will lean towards the gathering of water when an adult. If an adventurous thread awakens in her, she will be oriented towards the role of scout. The nymphs of these future scouts are identified by those who heat the hive and they benefit from an unusual treatment: they are incubated in higher-than-average temperatures which stimulate their intelligence and introduce them to thermodynamics.

When the young bee is born, she herself pierces the waxen capping which isolated her from the world. This *crossing of the veil* demands of her a determined effort, one which may endure several hours and which acts as a rite of passage towards incarnate life - an affirmation of her will to be born. Her arrival in the world is greeted with reverence and celebrated by an offering of sweet nectar. The benevolent caresses of the nurses daub her with royal balm and seal her membership in the hive.

The duration of gestation for maidens is 21 days, a figure of equilibrium *par excellence*. The 21 contains the wisdom of 1 – Unity – and that of 2 – Difference. The polarity of Spirit (+) joins with that of Matter (-). Drones remain in the cell 3 more days, thereby integrating more telluric ethers into themselves. As for the princesses, cosmic beings still incomplete at their birth, they stay only 16 days in their cradle."

A Cosmometric Genealogy

Among the strangenesses of the bee world, here is one of the most curious: parthenogenesis. This is asexual reproduction where an embryo can develop from an unfertilised female gamete. Among most species who practise this style of reproduction (like some reptiles and fish), the generated offspring is female. With the bees, however, when the queen deposits an unfertilised egg in a cell, the being to be born will be masculine. The drone receives half of the genes of his mother. He has no father, but he does have a grandfather. This particularity of the bee folk gives their family tree a unique cosmometry.

Human genealogy, like that of most animals, is constructed according to a geometric sequence with the common ratio of 2 (that is, each level is multiplied by 2). A child has two genetic parents, 4 grand-parents, 8 great-grand-parents, and so on.

If we observe the ancestry of a male bee, we see the appearance of a sequence of sacred proportions: 1, 1, 2, 3, 5, 8, 13, 21, 34… In this now-famous series known as the Fibonacci sequence, each term is the sum of the two preceding terms. The more we progress in the series, the more the ratio between two consecutive numbers approaches Phi, the Golden Ratio (1.618).

When the queen of a colony dies unexpectedly, the royal pheromone which inhibits the fertility of the maidens ceases to be diffused throughout the nest. In those cases where there is no fresh egg available and, therefore, she cannot be replaced by an 'emergency queen', a maiden then begins to lay eggs. Not having been fertilised, she will only produce male eggs. Even though the orphaned colony is condemned to die,[5] the genetic characteristics of her line will have the possibility of continuing to propagate, thanks to the males issuing from the laying maiden.

The family tree of the maiden differs slightly from that of the male: it starts one notch higher. The mother of the maiden will necessarily be a queen, but it may be that her paternal grandmother was also a maiden. Does this make sense?

---

5   If the orphanhood of the hive is discovered in time, an intervention may be possible. We would import a frame from another hive that contains freshly laid eggs, which offers to the colony the possibility of making a new queen.

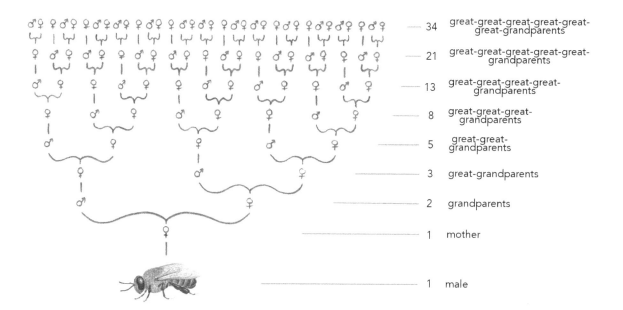

— 34	great-great-great-great-great-great-grandparents
— 21	great-great-great-great-great-grandparents
— 13	great-great-great-great-grandparents
— 8	great-great-great-grandparents
— 5	great-great-grandparents
— 3	great-grandparents
— 2	grandparents
— 1	mother
— 1	male

Genealogical tree of the drone

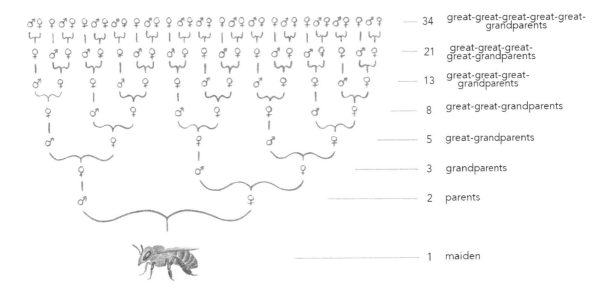

	great-great-great-great-great-grandparents
34	great-great-great-great-great-grandparents
21	great-great-great-great-grandparents
13	great-great-great-grandparents
8	great-great-grandparents
5	great-grandparents
3	grandparents
2	parents
1	maiden

Genealogical tree of the maiden

9
WAX: A MALLEABLE SKELETON

Translucent flakes shape the bones of our home,
Wax, solid and supple, is our structure of comb,
Archive of treasures, our wisdom of gold.

Encoded Wax

Friday, 21 December

On this solstice day, a number of my friends from the *BeeWisdom* network[1] are scattered in various spots around the globe. I have issued an invitation to a collective synchronised meditation, in order to enjoy together this important moment of transition and to dissolve the distance which separates us. The invitation spreads through the network, my friends invite their friends, and, rapidly, more than fifty people would like to join the experience.

I ask participants to procure a beeswax candle of good quality. As well as its exquisite odour and the sweetness which is exhaled while burning, beeswax warms the heart and reinforces our interconnection. Its physical and energetic structure allows it to conduct energy and information in a remarkable manner, so we will 'encode' our candle with a collective intention that will be freed and diffused in the noosphere during its combustion.

6:00 pm in Portugal, 11:00 am in northwestern Canada, 7:00 pm in Switzerland, and 3:00 pm in Brazil: participants settle themselves before their candle and connect to the network with heart and spirit. At 6:10 pm exactly, each proclaims the collective intention and lights their candle:

"Dear Bee Deva, I greet you.
Thank you for your wisdom, your love, and your devotion.
Thank you for your work as pollinator and as guardian of the Earth.
Thank you for your trust and your patience toward human beings.
As I light this candle, I invite responsibility, respect, and mutual support.
I invite healing of the relationship between Humanity and Bee.
So be it.
Aho."

My candle blazes up and crackles and I visualise a luminous constellation igniting itself upon the surface of the globe. A warm and intoxicating wave washes through my body. My heart swells with a profound feeling of trust and peace.

---

1   The *BeeWisdom* network will be presented in Chapter 22.

An Architectural Wonder

The magic which emanates from a comb of wax is enchanting... From time immemorial, the architectural talents which bees deploy to build their nests have fascinated humans. The harmonious structure of honeycomb is one of the most tangible manifestations of their intuitive knowledge of the cosmometry of the Universe.

A substance that is both strong and malleable, wax has extraordinarily plastic qualities. Thus, only 40 grams of wax, skilfully sculpted into hexagonal cells of comb, are capable of holding 2 heavy kilograms of honey.

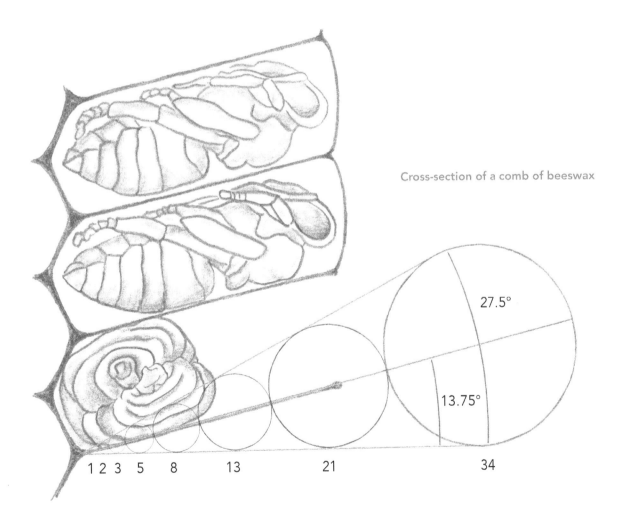

Cross-section of a comb of beeswax

27.5°

13.75°

1 2 3　5　8　　13　　　21　　　　34

In conventional beekeeping, the orientation of the combs and the size of the cells are imposed by pre-moulded wax foundations fixed in interchangeable wooden frames. When we explore the nests of colonies that had the opportunity to built in a natural manner, we observe that the structure of their skeleton is quite different than the one of standardised beehives.

The honeycombs are constructed vertically, from top to bottom, and are adapted to the cavity wherein the colony is installed. The cells are of varying size and depth according to their destiny. When we slice a comb vertically, we notice that the cells are slightly sloping upwards. This prevents the nectar from dripping when freshly foraged, and invites the brood to look towards the sky. While it may vary by a few degrees, the angle by which the cells are inclined is around 13.75°; this is half of the *Phi scaling angle*, obtained when tracing a series of circles whose dimensions, when placed tangentially, increase according to the Fibonacci sequence, as illustrated in the image.

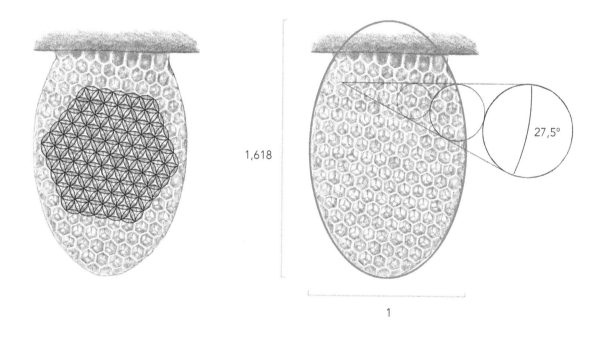

Flower of life and golden number in fresh comb

The constructions of fresh wax are as white as snow, and fragile. To augment their rigidity and solidity, they will then be covered with thin layers of propolis (this is what gives the candles their intoxicating odour). To begin, the baby combs have an elliptical shape, then the bottom parts enlarge. When held up to the light, we can see appear a transparent motif of the flower of life. The ellipses are generally built according to the *golden ratio* (ratio 1: φ in the example pictured here).[2]

All the naturally-produced combs that I have been able to observe until now display a slightly oblique hexagonal lattice. In numerous cases, we find again our Phi scaling angle between the horizontal plane and one of the upper faces of the hexagon.

In a natural nest, the tongues of waxen comb frequently undulate in parallel and a sensitive being may be able to perceive that their wavy forms emanate a profound softness. The interstitial space which separates the combs is finely calibrated, permitting bees to comfortably pass back-to-back and an efficient circulation of the internal air. The corridors should be neither too wide – to prevent drafts and heat loss – nor too narrow – so as not to suffocate.

The Voice of the Bee

"Wax is the mortar of our incarnation. It is the primary fundamental material of our structure, both physical and vibrational. It is dear and valuable to us.

The Hive Being is the architect who orchestrates the construction work of our wax skeleton. Her instructions resonate through the nest's information network, recruiting the necessary workforce for the construction site and transferring the plans to them.

Wax chain and catenary arc

---

2   There is a detailed study on the presence of the Golden Ratio in the dimensions of elliptical comb by Daniel Favre, 'Golden Ratio in the Elliptical Honeycomb' (2016); it is available online on the website of the *Journal of Nature and Science*.

When we are free to build according to our own criteria for harmony and efficiency, the orientation, size, and shape of the combs are chosen with logic and precision. It is not a mental and conscious choice; we follow the guidance of our collective intelligence. The orientation of the combs is aligned with certain telluric lines and adjusted to the geobiological conditions of the nest's location.

The physical construction of the combs is preceded by an astral outline formed of *wax chains*: the builders link to each other with the hooks on their legs. The chains that are formed draw the curves of the combs to be built. In this manner, the architectural plan is impressed upon the ether and serves as a template during the masonry phase. The wax chain helps the idea to descend into form. The information contained in the architectural plan is thus channelled and then crystallised in the structure of the honeycomb."[3]

Magic of the Hexagon

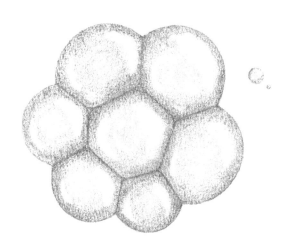

Seven soap bubbles juxtaposed

The builders produce the wax thanks to wax glands which develop below their abdomen. The previous painting shows a builder at work, adorned with 8 scales of freshly produced wax. The fresh wax is kneaded and moulded into cells with extra-thin walls that are simultaneously permeable and insulating. The cells under construction are initially tubular, then their molecular structure is modified to make the hexagonal form appear: by elevating the temperature, their plasticity is augmented and they naturally adopt a hexagonal arrangement, as do seven soap bubbles when juxtaposed.

---

3   This type of curve is called a 'catenary arc' (Latin *catena:* chain). It is largely utilised by human architecture in the construction of bridges, domes, porches, electrical networks...

Geodesic dome

The hexagonal form is absolutely fascinating. It possesses great intrinsic stability and we find it at every scale of our Universe, from the structure of carbon fibre to cloud formation above the north pole of the planet Saturn. It gives flexibility and solidity to certain bone tissue and to the protein membrane of our red blood cells. At the time of writing, I am fortunate to have lived in a geodesic dome for almost five years; a semi-sphere made of chestnut wood, wool, and canvas, the structure is composed mostly of hexagons (with a few pentagons). It is the softest and most harmonious space in which I have ever lived. Fundamental to my healing process, I can feel how the *medicine* of the hexagon protects me, centres me, and inspires me on a daily basis...

The 'honeycomb conjecture', now proven by Thomas C. Hales in 1999, states that 'a regular hexagonal grid or honeycomb is the best way to divide a surface into regions of equal areas with the least total perimeter'. As a storage space, then, a lattice of hexagonal tubes is the most efficient arrangement that exists – the construction of which demands the least amount of wax possible. But the charm of the hexagon goes far beyond its practical aspect...

The Voice of the Bee

"The hexagon reflects the perfection of Source in matter. Like the 6-pointed star, it represents the encounter between Earth and the Cosmos, the balanced union of Matter and Spirit. It expresses the sacred marriage of polarities that gives birth to form. Our wax combs behave like relay stations which disperse information to our surroundings.

The hexagonal pattern diffuses its *medicine* through numerous structural elements in the living world. In solid form, it manifests in crystals, clay, or snowflakes. In liquid form, we find it in *precious waters*[4] in abundance, such as the subterranean waters with which flowers make their nectar and the rainwater from a thunderstorm.

The wisdom of a hexagon has accompanied the evolution of human arts across the ages. Present in the flint of your prehistoric tools, the clay of your potteries, the glass of your windows and the optic fibres of your computers, it is a valuable ally in communication.

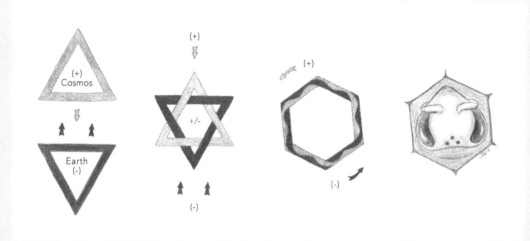

Union of Earth and Cosmos in the bosom of the cell

4   These 'precious waters' are called 'structured waters' in science. The vitality of water may be measured by Raman-Laser spectrometry, which makes it possible to estimate the presence of different types of polymers (molecular agglomerates established by hydrogen bonds) in water. It has been discovered that well-known healing waters and water from sacred sites (such as the water of Lourdes) contain more trimer-type polymers (molecules grouped in threes) and hexamer-type polymers (grouped in sixes), and, particularly, hexamers of hexagonal shape. Source: Jacques B. Boislève, *Structure et propriétés de l'eau*, 2010.

Whether they are microscopic, mesoscopic, or macroscopic, all hexagonal structures of the Universe are in resonance with each other. They form an immense communication network which participates in the expansion of universal consciousness. This network anchors more dynamic co-creative forces induced by the lemniscate. Hexagon and lemniscate thereby act in synergy to construct and anchor the world.

When the descending energy of the Cosmos and the ascending energy of the Earth meet in the wax cell, they love each other, they embrace and melt into each other… The brood which sojourns there are nourished by this hug; they vibrate and lean from the songs which circulate through the universal hexagon network. Perhaps one time, if you feel like it, you will let a drop of honey – just one! - dissolve slowly, softly, on your tongue. You will close your eyes…. And you will hear it: the song of the hexagon… Then you will feel the deep trust that reigns in this crystalline cradle. And you will allow a profound joy to settle within…

The wax is a reservoir of *concentrated light* that shines in the darkness of the hive. A kilogram of wax is the concentration of luminous energy from 9 kilos of honey, itself the concentration of luminous energy from 45 kilos of nectar, which is composed of water and the sugars which carry within themselves a crystalline structure. Be aware that this energetic chain is released when you light a candle; the energetic difference from burning a paraffin candle is obvious. Beeswax's qualities of inspiration, protection, anchoring, and conducting information are greatly beneficial to human meditations and prayers.

In the womb of the hive, the wax comb constitutes our *library*. All the history of the colony is archived there; it contains the valuable database of pollens, plants, minerals, and other biological characteristics of our environment. Memories of the chemical treatments administered by the beekeeper are also preserved there for many years. Leaves of pre-formed wax brought into the hive by the beekeeper, made from a mixture of melted waxes, make us dizzy. Imagine if you tried to read a book where the words were melted and mixed!"

10
HONEY: LIQUID GOLD

Mouth to mouth, the nectar is ripened,
Refined into vital abundance and glee,
Laid down in cells by sweet alchemy.

Solstice honey bath

Wednesday, 21 June

To celebrate the summer solstice, I have invited some friends to share a unique experience in company with the bees: a honey massage. Here we are, all six of us settled near *Inanna*, one of the hives close to my house. A little bit of vegetable oil is added to honey and we apply this precious balm to our skin. Eyes closed, lulled by the prosperous hum of Inanna, we listen to the pores of our skin delicately opening under the gentle and penetrating effect of honey. Now and then, a few bees circle around us, but the nearby fields are so engorged with fresh nectar that they prefer to orient their attention towards the flowers. The golden liquid honey cleans us and nourishes us with its richness. The sensation is captivating, even erotic. The session terminates with bursts of laughter under the jet of a garden hose.

Gift of Heaven

...When Golden Suns appear,
And under Earth have driv'n the Winter Year:
The winged Nation wanders thro' the Skies,
And o're the Plains, and shady Forrest flies:
Then stooping on the Meads and leafy Bow'rs;
They skim the Floods, and sip the purple Flow'rs.
Exalted hence, and drunk with secret Joy,
Their young Succession all their Cares employ:
They breed, they brood, instruct and educate,
And make Provision for the future State:
They work their waxen Lodgings in their Hives,
And labour Honey to sustain their Lives.
The Gifts of Heav'n my foll'wing Song pursues,
Aerial Honey, and Ambrosial Dews.
Read this other part, that sings
A mighty Pomp, tho' made of little Things.

Virgil, selections from Book IV, *The Georgics*, 30 BCE
translation by John Dryden, 1697 CE

I like this poem, which sings the magic of the little things... Since forever, honey has fascinated humanity with its evocation of joy, sweetness, and eroticism. Venerated for its physical and spiritual qualities, honey has accompanied human history since the dawn of time as a symbol of protection, of abundance, and of eternity.

Among the numerous properties of honey, I am struck by its *adaptogenic* quality. It is in some way a living being in its own right, one who *knows* how to adapt to the needs of what it touches. For example, honey is both antiseptic, destroying pathogenic bacteria, and prebiotic, stimulating the growth of beneficial bacteria in the intestines. It cleans and nourishes simultaneously, dissolves and reassembles, inhales and exhales...

In the human mind, the bee is mechanically associated with honey production. When I mention that I work with bees to people who don't know me, one of the first automatic questions that I receive is often "Ah, so you make honey?" I smile. It is the bees who make the honey, and I help them as best I can. Above all, though, I become intoxicated with the invisible honey of their singing and dancing. The physical honey harvest has never attracted me, but occasionally it happens that they offer some to me. I use it as a medicine, a valuable gift, or a celebration.

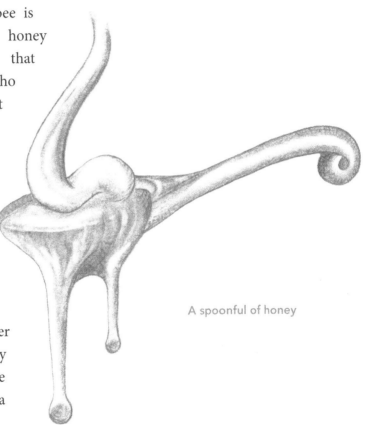

A spoonful of honey

The general situation of the global honey industry reflects, in a conspicuous manner, the irresponsible phase that humanity is going through. A large portion of the honey for sale through mass-market retailing – almost 2 million tons per year – is adulterated. Conventional apiculture is made up of practices that are, much of the time, very disconnected from natural laws. Consider, for instance, sugar feeding. I invite you to put yourself in the place of colony for a moment: you fabricate one of the most exquisite and elaborate substances in the world, and a giant creature in a cosmonaut suit comes and takes it away, and replaces it with sugar candy…

White sugar, symbol of the modern capitalist system, provokes similar effects in humans and bees. First, it overexcites, then it gives the *blues*, inciting more consumption. *The vibratory signature of refined sugar puts us in a frenzy*, the Bee Deva tells me. *When colonies are fed with it, their connection to the Bee Deva and with the heart of Gaia is enfeebled. These colonies become more reactive and may develop a certain animosity towards the human being. They are also more vulnerable to illness.*

Today, addictive behaviours and mood swings linked with human sugar consumption are more and more recognised. As a consequence, many people consume honey as a sugar substitute, imagining they are doing better. Unfortunately, the hives from which this honey was extracted were often themselves fed with sugar.

Some bee protectors claim that it would be better to cease all honey harvesting, across the board, in order to 'save' the bee. It is obvious that beekeeping practices and the behaviour of consumers are called to change radically. It is encouraging to observe that awareness of the bee situation and the quality of honey is growing; television shows, documentaries, internet articles, conferences and literature on the subject are abundant. The world is changing…

When a relationship from soul to soul is established between Bee and Human, it is – in my opinion – possible to blend these approaches in a manner that is entirely respectful towards the bee Being. Some hives may be devoted to offering honey surpluses if a sufficient number of other hives are left untouched. A colony in good health is capable of producing more honey than she needs. When treated with reverence, a prosperous colony may offer part of her treasure with joy, in the same manner that healthy and balanced human beings like to offer their gifts to those around them. Prosperity and abundance go hand in hand. Honey is not something *to take*, but, rather, *to receive* - as an offering, with gratitude and with humility. The quantity of honey collected in this manner is much less than the average harvests of conventional apiculture, but this honey is of an entirely different quality. The properties of honey gathered with the blessing of the Hive Being are profoundly beneficial and therapeutic.

Honey is not to be consumed as a substitute for granulated sugar, but as a 'heavenly gift', delicate and precious. Therefore, those who love the bee will make of its consumption a conscious choice. Dear readers, if you decide to make this choice, I invite you to be discerning about the quality of this sweet offering and, as best as possible, to procure for yourself local honey that is raw (unpasteurised), unfiltered, and unadulterated, and which comes from wildflowers or organic crops. To invest a few extra euros in a jar of honey generally agreed to be of a superior quality. Know, as well, that the enzymatic richness of honey, its vitamins and organic acids are extremely fragile and easily deteriorated by numerous factors, such as heat, exposure to light, contact with air or metal, electromagnetic waves, barcodes, and environmental stress.

The Voice of the Bee

"The fruit of concerted efforts from the mineral, vegetal, and animal realms, honey hums the song of abundance and generosity. It is the outcome of a long process of alchemical maturation across the kingdoms. The water with which flowers manufacture their nectar comes from the bowels of the Earth, radiating its *black light*. Charged with *mineral wisdom* from the Earth Mother, this vibrant water is sipped up by plant roots and enriched with sugars, amino acids, and essential oils. During this process, the water absorbs specific properties from the plant into which she is welcomed. Thus, *vegetal wisdom* is added to mineral wisdom. The refined nectar is preserved in the womb of the flower, in the core of its chalice. It is the feminine erotic liquid of the plant which is offered to bees, an exquisite compensation for their precious services.

Once brought to the hive, the nectar is *sublimed*[1] by the work of our alchemists. Our mouths and our crops[2] act like alembic flasks which distil and condense the devic essence of nectar. The specific enzymes that we combine for use are like thousands of little flames which melt, remelt, and refine the nectar's molecular structure. It goes from mouth to mouth, infused with the energy of giving and receiving. Then it is stored in cells and concentrated by the fanners. A few grains of pollen are dropped into it, a white touch of Yang fertilising the black of Yin.[3]

Honey *sticks*, in every sense of the word; it is a binding agent which re-unites both physically and spiritually. Honey is the *guardian of heat*. At the physical level it is our heating fuel, making an insulating overcoat which protects the colony from winter cold. At the social level it guarantees communal heat and solidarity. Its message is one of sharing; it recalls the unity of the world beyond space and time. The love which it radiates seals the coherence of existence. Whoever knows how to listen to the song of honey will never feel alone."

---

1   To sublime: to render finer; to convert (something inferior) into something of higher worth.
2   In bee physiology, the crop is also known as the 'honey stomach' or 'honey sack'. The main function of the crop is transport and storage of liquid food (water, nectar, or honeydew).
3   A reference to the symbol of the Tao, where the dot of pollen/yang/white gives life to honey/yin/black, and vice-versa. Nectar circulates from mouth to mouth through trophallaxis, as illustrated in the preceding painting. Through the fanning process, moisture content falls from around eighty percent in nectar to seventeen percent in honey.

11
POLLEN: STRENGTH AND CLARITY

Seeds of stamens sublimed by enzymes,
Nuggets of pollen, the colour of rainbows,
Collect and sparkle like gold dust of stars.

A Pallet of Pellets

Sunday, 8 April

These last few weeks, I've been inventing a detective game: stationed beside a hive, I observe the foragers returning from their peregrinations loaded with multicoloured pellets. Mentally, I note the different colours which file past – from golden yellow to spiced ochre, from pale orange to deep carmine, from veined white to arresting purple... Then, while I ramble around the fields and forests, I try to connect them with their flowers of origin. This is not such an easy exercise. The colour of the pellet does not always correspond to that of the flower, and for several among them, my internal investigator is challenged. The flower-pellet link reveals itself with certainty when I come upon a forager in the middle of collecting.

As I write, there is an abundance of *cistus*, great suppliers of a sparkling orange pollen, which contrasts with the pastel blue pellets from wild lavender. From time to time a rare colour appears, such as this grey so dark as to be almost black – perhaps issued from a poppy flower in a neighbouring garden? - or this vermilion red, of which I could not find a trace...

I find that my exclusive reliance upon the sense of sight for performing this exercise is very limiting. The bees, now, they use a much more diverse palette of senses. They recognise the nectar and the pollen of flowers by their odours, their vibrations, and their songs, which they have heard and tasted from their youngest days in the heart of the hive. Even though I specialised in botany for several years and learned to attribute complicated names to the plants, I realise that the bee has knowledge about plants which is much greater and more precise than mine. I smile with humility...

Full Baskets

The health of bees is intimately linked to that of the plants and, above all, to the diversity and quality of pollen sources to which they have access. As the colony's essential nutrient base, pollen secures her growth, her strength, and her immunity.

Seventy per cent of flowering plants are pollinated by insects; this partnership is one of the most fundamental to the equilibrium of our ecosystem. During more than 100,000 years of co-evolution, each species has developed unique morphological

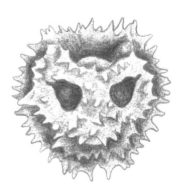

A grain of dandelion pollen enlarged through an electron microscope

and behavioural characteristics. For example, when observed under an electron microscope, grains of pollen present the most astonishing shapes. The above drawing shows a grain of dandelion pollen magnified several hundred times. The grain is ornamented with a multitude of small spikes which allow it to easily hook onto the forager's hair. The bee will serve him as a taxi to pay court to female gametes in other flowers of the same species.

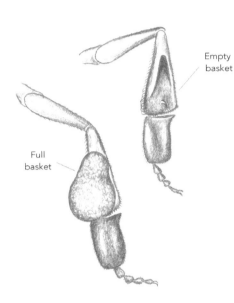

Empty basket

Full basket

Rear leg of a maiden

To observe a bee collecting pollen makes me smile with amusement and admiration. The precision with which she uses her harvesting tools is impressive: "*Scritchscritchscritch*, I comb the stamens, *rubrubrubrub*, I gather the grains into a ball and soak them with saliva, *patpatpatpat*, I pack them into my little baskets, and *bzzzzzzzz...* I continue my rounds!" And this carries on for hours...

The rear legs of the forager are hollowed into the form of a basket, wherein they bear their harvest during flight. Having been coagulated into pellets, the grains of pollen are solidly stowed there thanks to a tiny curved spur - a marvel of engineering!

The Voice of the Bee

"A grain of pollen is a spark of light. It is a conducting vector for the masculine principle which descends from the Cosmos to inseminate the Earth, inviting her to bear her fruit.

Pollen sings the beauty of the landscape: its subsoil, its mantle of greenery, its climate, and its character. The pranic energy[1] which emanates from pollen is a fundamental source of nourishment for the colony and, above all, for the brood. When the larvae are fed with it, they discover in the cradle that which awaits them outside. They are imbued with the cartography of their territory. When they later become foragers, it is with joy that they find themselves in the belly of flowers whose flavours, odours, and vibrations nourished their growth.

After the pollen is brought back to the hive, it is, like nectar, sublimed[2] by the alchemists. The raw grains of pollen are equipped with a protective shell and when we open these insulating shells, the message of the pollen releases itself and sings through the hive. This is the role of the handlers, who use enzymes, bacteria, and yeast to liberate the devic essence of the pollen and to render it bio-assimilable. The pollen is then stored in cells with a little bit of honey,[3] the black drop of Yin fertilising the white of Yang. Like carbon which transforms into diamonds in the depths of the Earth, raw pollen becomes *sublime pollen.*

Besides the geological, geographical, and biophysical information which it transfers, pollen is a source of many other teachings for the bee people. The intrinsic quality of pollen as the male gametes of the plant people is to stimulate strength and responsibility, clarity and passion. The ardour of the maidens at work is boosted by this powerful fuel. It teaches discipline and generates initiative, creativity in action. *It brings the future into the present.* 'Shoulds' do not exist; devotion to placing oneself in service to the common good is natural and spontaneous. Pollen invites us to live intensely in the present, passionate about life, with patience and discernment. Strength and softness combine, carry us, guide us, and delight us..."

---

1   *Prâna* (from the Sanskrit *prâ* – constant – and *na* – movement) is the original energy of life which infuses and manifests itself through every object, animate or inanimate.

2   For the definition of *to sublime*, see Chapter 10, note 1.

3   On the previous painting, a maiden handling the pollen uses her head to pack it in the bottom of a cell. The process of anaerobic lacto-fermentation which is then produced within the cell renders the pollen much richer, more nourishing and assimilable by an organism (human as well as bee). This is commonly known as 'bee bread'; personally, I prefer to call it 'pollen bread', or, as does the Bee Deva, 'sublime pollen'. If you consume pollen which you have purchased in its dry state, *open it* before ingestion by leaving it to soak for a minimum of one hour in a little honey water or lemon water.

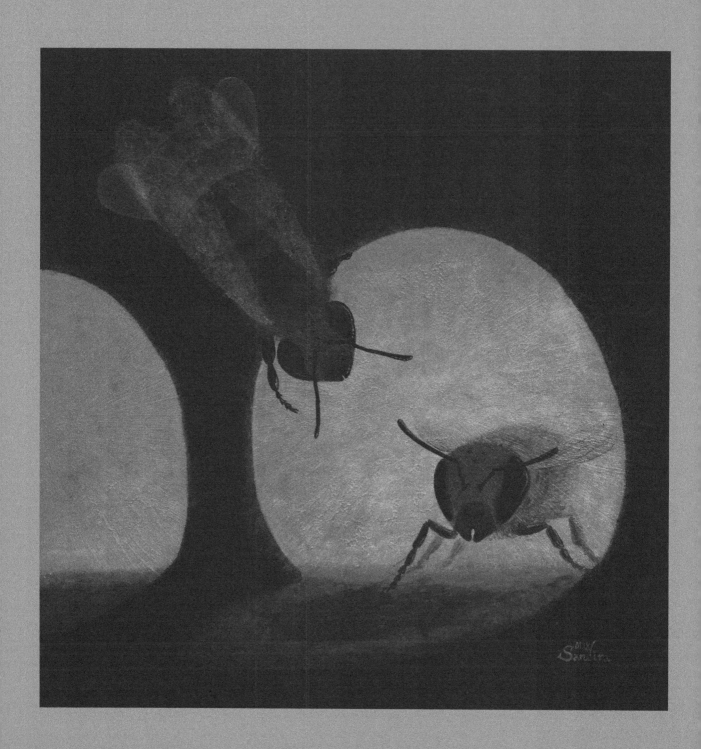

12

PROPOLIS: INTERFACE BETWEEN WORLDS

Well-guarded steps of the sacred temple,
Keyhole between the light and the dark,
Inner worlds and outer, connected together.

Mummified Alive

Thursday, 25 June

This morning, we are moving the hive *Choukran* ('thank you' in Arabic) into a brand-new hive body. At the bottom of the old box, there is a curious sight: several little hills of propolis strewn across the floor and some of them are *moving*!

These piles of resin are actually the sarcophagi of black beetles.[1] Smothered with propolis, a few among them are still alive and wave their feet in a vain attempt to free themselves. Great lovers of honey, these intruders are abundant this year and because they are protected by their thick shells, the bees are unable to get rid of them by stinging. Sometimes, though, they manage to immobilise them and mummify them with propolis.

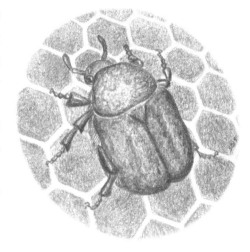

Black beetle on a ray of comb

Black beetle mummified in propolis

Particularly rich in propolis, this colony has also sculpted it to adorn their entrance with magnificent propolis-vaulted porches (as illustrated in the preceding painting). I admire the talent of these masons. A tip of the hat, lovely maidens!

---

1   Native to Africa and about 2 cm long, this chubby and greedy beetle (*Oplostomus fuligineus*) likes to set up house in the abundant pantry of a hive of bees.

The Gates to the City

Harvesting the resin from a poplar bud

The word *propolis* comes from Greek, meaning *pro* – for or before – and *polis* – the city. Physically and symbolically, propolis marks the borders of the hive-temple. This boundary is not an airtight shield, but a thin skin which is simultaneously highly protective and breathable. The Propolis Deva is the guardian of the threshold between the worlds within and without, the point of contact between shadow and light, night and day. If propolis was a moment of the day, it would be the twilight.

Propolis is composed of roughly 50% resin collected from nearby plants, 30% wax, 10% pollen and 10% essential oils. It is fundamental to the maintenance of the colony's immune system and it is an adaptogenic antibiotic. That is to say, it does not eliminate everything in its way like an artificial antibiotic, but *knows* how to discern between beneficial and harmful substances.

My medicine chest for travelling is quite minimalistic but it always has propolis, which adapts to many situations. It can be chewed pure, like chewing-gum, and is effective against aphtha as well as against gum infection. It can be used as a temporary filling to fill a dental cavity and to ease pain. It supports digestion and calms sore throats. For this last, I prefer a salted extract: I reduce the propolis to powder and let it soak for a lunar cycle in a hypertonic salt solution. Some like it in an alcoholic tincture to treat wounds or sores, or even in cream with a beeswax base.[2]

---

2   Hypertonic saltwater contains the same proportion of salt as sea water: 35 grams per litre.

The Voice of the Bee

"Propolis is our *skin*. It is our protective membrane and our contact with the outer world. All the interior surfaces of the hive, both of wood and of wax, are covered with a coat of propolis in greater or lesser thicknesses. It serves to fill in any cracks and to harmonise the circulation of energy, as well as preventing draughts and the development of pathogens.

The ingredients which we assemble for its formation are meticulously measured. The resins are harvested from our plant allies, particularly the trees, who offer it to us with kindness. Each plant spirit bestows its particular properties upon its resins. We know how to recognise them, how to differentiate between them, and how to assemble them to create specific remedies which are adapted to the particular health challenges that we face. We prefer to have a large diversity of resources available. The resins of indigenous plants are precious because they have adjusted to our needs during the course of our common history.

When a colony is sick or out of balance, her alchemists go to work devising the most effective remedy for self-healing. As well-intentioned as it may be, opening the hive during this fragile state is often more destructive than helpful. The human being will help us more with their benevolent presence and the amelioration of environmental conditions around us than by their intrusion into the nest.

Propolis acts as a filter of electromagnetic, telluric, and cosmic radiation. When the *propolian* envelope is pierced, we immediately sense it - just as your organism feels when some object pierces or damages your epidermis. Each time a human opens the hive, the propolis seal is broken. If the opening is sudden and without a preliminary contact with the Hive Being, the shock may be brutal, especially if the colony is fragile. Cosmic, climatic, and environmental influences instantly penetrate the body of the hive. If the exterior conditions are favourable, the impact is less. However, if the air is humid, the moon is at a node, or the beekeeper is nervous, the ensuing repair work demands some serious effort."

Protection?

Sunday, 11 May

I am participating in a three-day workshop on organic apiculture being held in the eastern Alentejo region of Portugal, close to Mértola. I know I will not learn a great deal theoretically, but my intention is to connect with the Portuguese beekeeping community and to gauge its openness towards natural beekeeping practices. And since my Portuguese is not very fluent yet, it's a good opportunity to expand my vocabulary. There are a dozen participants: 2 women and 10 men.

The third day is devoted to a field trip visiting the beeyard of a local beekeeper. In the car park where we assemble before going to the beeyard, each participant is urged to don a full bee suit. The energy of Venom and the fear which it awakens are in the air. A sort of tense laughter attempts to disguise the general stress of the group; jokes are fired off about who will be stung first. Some of the people wind several layers of tape around their ankles and wrists in case any 'vicious' bee becomes entangled in a sleeve.

We arrive at the beeyard, where some thirty grey boxes are lined up in tidy rows; each one is identical to its neighbour. We look like a group of cosmonauts on a moon mission. The beekeeper opens up the top of a hive like he was opening the hood of a car. His speech is mechanical and he talks as though the hive were a machine, describing how to recognise each piece, its form and its function. He does not seem to realise that he is opening up a living being.

When he lifts out a frame of brood, the colony flares up; a storm of angry bees breaks over the congregated group. However, nobody reacts and the professor continues his demonstration as though nothing was happening. Under their thick armour, some of the people seem to appreciate the experience of feeling sheltered from the venomous arrows.

I remain somewhat in the background, attempting to calm my inner irritation with this attitude so disconnected from the Hive Being. The sun is beating down and my body is stifling under the suit, so I pull off my gloves to let my skin breathe and to let the bees know that there is a body of flesh and bone beneath this defensive envelope. Even though not a single bee is interested in stinging my fingers, when the beekeeper sees me, he is startled and immediately urges me to 'come to reason'...

I learned a great deal during this course. I understood all the more strongly the manner in which mechanical man dissects the living world like a machine. I found myself even more determined to propose a sensitive and holistic approach for those who seek a different way of connecting with the world.

One of the reflections which emerged from this experience concerns the notion of 'protection'. As I write these lines (October 2018), it is rare for me to use a veil when visiting my hives. This was not always the case. When I began, the security afforded by a veil as I learned to enter into contact with the bees was valuable to me. I learned to *feel* the Hive Being, her mood, her overtures, her warnings. I always have a veil with me in case I need it, and for some procedures, it is a true ally. When I bring a visitor to the hive, I ask them to feel inwardly what is true for them at this moment, independently of all self-judgment or preconceived ideas. I specify that this decision is their responsibility, not mine. I make available different types of veils covering the top half of the body or just the head. To build a relationship of trust with the Hive Being, it is essential to feel safe, while remaining open to constantly adjusting behaviour, so that intimacy may blossom and be refined.

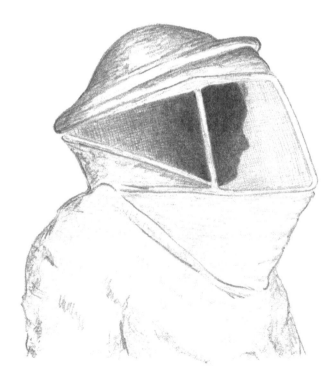

Beyond the veil

The Voice of the Bee

"Propolis is our veil of protection. *It holds us*.

It forms a breathable garment which secures our integrity and defines our contours. This envelope is not an armour; its permeability allows us to remain in contact with our environment even while protecting us with its reassuring presence. Day and night, we bathe in its odour and its vibration, which gathers us and embraces us.

To know how to protect oneself without closing down is an art.

When you make the decision to use a veil when visiting a hive, know what motivates your decision. Are you carrying the veil like a warrior leaving for combat carries a shield? Does it function as an impenetrable mask intended to render you insensitive to genuine contact? Or do you use it as an intermediate tool which ensures and supports the establishment of a climate of trust and sharing? If you do not know exactly how to answer these questions, observe the response of the bees to your approach. Mirrors of your inner reality, they will give you valuable feedback for getting to know yourself and for making conscious choices for your evolution."

13
THE FORAGERS' BALLET

Drunk with voluptuous perfumes and colours,
I offer my kiss, my caress, to the flower,
In sensual bliss, dancing together.

Phoenix Reborn from the Ashes

In June 2017, the centre of Portugal was devoured by enormous wildfires, among the largest and most deadly of its history. Several tens of thousands of hectares of land and forest were consumed and hundreds of people found themselves homeless. Beyond its environmental impact, the fire has touched hearts and minds. The country's forest management system, essentially based on the intensive production of eucalyptus, must be rethought from top to bottom. These monocultures are deserts which erode the soil, sterilise the earth, and catch fire from the smallest spark like a box of matches.

Situated at the heart of this region, Gravito is a magical and bountiful small hamlet which hosts seminars and retreats. I have worked for several years in collaboration with my friends Miguel and Shobha, who guide and cultivate the centre with passion and devotion. The flames razed their land along with the yurts and tipis which welcomed visitors, but their house was miraculously preserved from harm. The hive which we had installed together was reduced to cinders and smoke. I visited them three months after the fire and we decided to bring over a young colony of bees from my place, in order to support the regeneration of the burnt earth.

Friday, 27 October

Awake at 4 am. After fitting the entrance with wire mesh, my friend Annelieke and I load the hive *Flower of Life* into the boot. The Bee Deva advises me to orient the entrance of the hive towards the front of the car. Just like human beings, facing in the direction of travel means that it takes less energy from the bees to adjust to movement. We arrive at Gravito around 10 am; it is beginning to be hot in the car and the bees are impatient to get out. We place the hive in a section of the garden that was untouched by flame.

We are both curious and a bit worried. How will the bees adapt to this charred landscape? Will they find the necessary means to feed themselves? Would it be best to feed them or to move the hive to the neighbour's lands (which were not burned) during the winter months? For the moment, they do not show any sign of aggression and immediately begin to explore their new environment.

I am surprised by the vitality which emanates from the landscape. It has not rained yet and, even so, the earth and the burned trees are covered with young shoots in tones of green that are bright and varied, almost luminescent. It is like the dawn of a new spring. Few birds and insects survived and a deafening silence emanates from the forest. In the depths of silence, I perceive the murmur of the Earth being reborn and the song of a phoenix emerging from its ashes... Where I anticipated enduring the blow of a dead and desolate landscape, I am welcomed by an oasis of burgeoning life.

Monday, 30 October

Three days after we arrive, the hive is showing an impressive level of activity. The flight corridors of the foragers form aerial tentacles around the hive; the foragers, full of nectar, stagger under the weight of their booty when landing upon the entry board. The density of traffic and the routes of the aerial corridors vary during the course of the day. In this landscape which appears empty of flowers, where do they find their harvest? During my walks around the property, I sharpen my senses. This morning, at dawn, I am called by a powerful buzzing emerging from a great oak. As I approach, I witness hundreds of bees greedily licking the glistening leaves. Curious, I do the same and discover that the leaves are sweet![1]

Later in the day, I sit myself down beside *Flower of Life* and connect with the Bee Deva. She explains to me that during the regeneration phase of the Earth after a fire, the presence of bees has a healing and revitalising effect on her energy systems. Once more, I understand how much the bee is a model of devotion and service...

Expert Pollinators

When a bee strokes a flower and absorbs its sweet beverage, the blossom also sprinkles her with pollen dust. Our ambassador of love transports this to the next flower. With neither wings nor feet with which to move, flowers need a mediator to accomplish their love-making. The majority of flowering plants have chosen pollinating insects as vehicles for their male seeds to fertilise the female ovaries hidden within the pistil. Insects are much more efficient than the changing winds, and among them, bees are particularly prized. Indeed, their services offer a unique expertise. On the one hand, unlike butterflies (who flutter about here and there), bees are loyal. Once they begin to forage from a certain kind of flower, they stick to it with fervent constancy until the source of nectar is exhausted. This behaviour works well for the plants, whose desire is to see their pollen spread among other flowers of the same species. As well, the spectrum of plants which is visited by the bee is particularly vast, contrary to the bumblebee, who is also extremely faithful but whose palette is smaller.

---

1   This viscous and sweet substance is honeydew and much appreciated by bees. It comes from the excreta of sucking insects (aphids or scale insects) which feed themselves with the nitrogenous products of sap and reject the sweet part.

Over the course of their evolution through the ages, flowering plants have refined their morphology to optimise their pollinating efficiency. The sage flower, for example (as illustrated in the preceding painting), has put in place a particularly judicious mechanism: its lower petal, an attractive runway for the foraging bee, works as a switch. When the bee settles herself on this 'rumble seat' to draw up the nectar, the stamens drop down and deposit their pollen on her wings. Several days later, this flower's stamens fade and it is the pistil which lowers itself this time, kissing the wings of a forager who carries pollen from a neighbouring flower...

The Erotic Body of Earth

I like to visualise the vegetal cloak of the earth as the erotic body of Gaia. As a matter of fact, among the various realms, the vegetal is the one which has most developed the art of seduction. Plants have deployed incredible artistic talents to seduce the insects carrying their precious semen.

The 'courtship ritual' of flowers for attracting pollinators is conveyed by a multitude of visual, olfactory and electromagnetic signals:

- Visual signals: the visual spectrum of the bee spans higher vibratory frequencies than our own; while red seems like black to her, she can clearly perceive ultraviolet. Thus, numerous flowers adorn their petals with various motifs which are invisible to our eyes but which indicate the path to the chalice for pollinators. Certain orchids imitate the form of the insect as an invitation for her to come and make love with them....

Bee orchid flower

- Olfactory signals: the bee possesses sixty thousand olfactory receptors concentrated on her antennae, which permit her to discern an almost unlimited number of odours. In order to detect the locality of fertile flowers, bees follow the trail of their fragrant discharge along the wind, a valuable ally.

- Electrical signals: the bee is charged with positively-charged particles (+). When a flower is at the peak of ripeness, it emits a light electric field that is negatively charged (-); this draws the bee towards it like a magnet. Then the signal disappears during the time needed for the flower to renew its supply of nectar. Concurrently, the forager places a chemical marker in the flower which informs her fellows that this bloom is temporarily 'indisposed'...

The olfactory and electrical signals indicate equally the flower's fertility status. Once, I happened to observe a bed of flowers strangely ignored by the bees for some time and then, overnight, covered in foragers. *Flowers enter into their fertility gradually*, the Bee Deva explained to me. *Our maidens know how to read this curve and, insofar as other sources of nectar are available, choose to forage among them when they are ripe, at the zenith of their fecundity.*

Go to your fields and gardens
And you shall learn that it is
the pleasure of the bee
to gather honey of the flower,
But it is also the pleasure of the flower
To yield its honey to the bee.
For to the bee a flower is a fountain of life,
And to the flower a bee is a messenger of love,
And to both, bee and flower,
the giving and the receiving of pleasure
is a need and an ecstasy.

Kahlil Gibran, in *The Prophet*, 1923

The Voice of the Bee

"Bees and flowers love each other immeasurably. For millennia, they have woven a relationship of trust and of co-operation which goes far beyond the simple practical function of pollination. When a foraging bee lands on a flower, she says to it '*I see you, I know you, I thank you for living*'. This gratitude is genuine nourishment for the etheric body of the planet.[2]

Gaia's vibratory structure is composed of a unified complex of energetic networks. These interlaced webs structure and maintain the Earth in a coherent unity. They may be compared to layers of finely woven garments which clothe and protect our planet. All the networks are interconnected and constitute a terrestrial *intranet* through which billions of pieces of information permanently circulate. By flying from flower to flower, the bees stimulate the energetic flow which circulates through the network.

Foraging bees cover thousands of kilometres through the air.[3] Their translucent wings behave like prisms which refract the light of the sun. Their movements are stamped into the ethers and leave behind a luminous vibratory imprint which endures some time before dissipating. During their journeys and explorations, they are permanently linked to their Hive Being by a sort of 'umbilical cord'. This powerful and elastic link helps them to find the path back to their mother, like Ariadne's thread in the labyrinth.

---

[2] Gratitude, love, and positive thoughts stimulate growth and prosperity. Like a flower, when a bee lands on my skin, I feel love nourishing me. This brings to mind the following experiment conducted by the Japanese researcher M. Emoto: three jars were filled with rice, which was covered with water and stored in the same conditions. Each day, the words 'Thank you' were spoken to the first jar, 'You are stupid!' to the second, and the third was ignored entirely. At the end of a month, the rice in the thanked jar, still white, had developed a beautiful and healthy lactic fermentation, the second became black and the third had rotted. (source: documentary *The Secret of Water*, 2015)

[3] According to the availability and quality of available food sources, bees can cover more than 100,000 kilometres to create one kilogram of honey, a distance equivalent to several times around the planet.

Physical sciences explain electric current as generated by a difference of potential. In the same way, in order for the *terrestrial erotic current* to circulate, there must be a difference of potential. During our flights from flower to flower, we stimulate the terrestrial Eros by transporting the electrical charges contained in pollen (+) and nectar (-). This activation of masculine and feminine principles is an *energetic pollination*; its effects are profoundly revitalizing. They propagate themselves in the vibratory field of the planet through her energetic networks.

The *complex simplicity* which regulates this mechanism secures for the planet her strength and her resilience. For all that, the energetic webs are fragile and subject to ripping. Actions and behaviours which are not in harmony with natural laws cause many wounds within the networks. When humans scrape the surface of the earth with machines, pour pesticides upon her, raze the forests, and kill the whales, when they play with nuclear fission, gene manipulation, and wireless technologies, when they unleash violence in wars and tyrannies, they pierce the energetic cloak of the Earth. Today, Gaia's clothing is in tatters.

Like devoted seamstresses, bees do their best to repair the rips. But when the weave is too damaged, the bees become exhausted and they need to renew themselves. When bees die or mysteriously disappear, they are withdrawing to other dimensions to rest and to recharge their batteries.[4]

To be strong and durable, this repair work requires collaboration with human beings. The bee is ready. She awaits your awakening."

---

4   This refers to 'Colony Collapse Disorder', a phenomenon observed more and more during the last few years. When CCD occurs, bees disappear from their hive overnight without leaving a trace, even though large stores of honey and pollen may remain in the combs. Many factors are given to explain this phenomenon, including pesticide use, the poverty of monocultures, and the development of wireless technology.

14
THE ARTS OF LANGUAGE

I speak the language of sun, wind and flowers,
I murmur and hum the atomic script,
My dance is a poem, tracing infinity.

Messenger Bee

Wednesday, 26 April

Gosh, I lost track of time!

My guests are arriving in ten minutes and I am not ready. I would like to welcome them with tea and a small homemade snack. Quick, quick! I busy myself in the kitchen, absorbed in my thoughts.

A low humming sound suddenly pulls me out of my bubble. A dozen bees have come in through my window and are actively wheeling around me! One of them positions herself in a stationary hover some twenty centimetres from my forehead. My first reaction is typically narrow-minded: *What are they doing there? Hey, I don't have time to deal with you!*

But the bee insists.

My second reaction does not fly much higher: *Bizarre... they must have been attracted by a sweet odour.... have I left a jar of honey half-open somewhere?*

The bee perseveres.

"Shhh.... Listen... she says.

– Listen? What?

– Go to Tiamat.

– But, uh... why?

– Go to Tiamat."

A hallucination? A spiritual fantasy? *I won't know unless I go.* Four minutes before the anticipated arrival of my visitors and my snack is still not ready... *Fine, I will go to Tiamat.*

Tiamat is a hive situated a few hundred metres from the house, in the middle of a thicket of rockrose shrubs. I draw nearer, curious... and I see it! Enormous and serenely suspended, the opulent cluster of a swarm. It vibrates and shines, solidly secured to the low branch of a small oak tree a few metres away from the mother hive. The moment is magical.

But here are my guests. *Back to the material.* All enthusiastic, I announce the new programme to them: to collect the swarm and install it in the *sun hive*, a straw skep hive that I built. I invite my visitors to assist with the process, at a respectable distance so as not to interfere energetically.[1] The swarm allows itself to be collected like a ripe fruit. The use of a protective veil is unnecessary, as trust obviously prevails.

A magnificent gift for each one of us. Thank you, beautiful bee!

---

1   The *sun hive* is a straw hive in the shape of an egg. For the bees, it is a real 5-star hotel. Its oval form is that of a natural swarm and the materials involved offer them outstanding comfort.

Sun hive

The new colony is installed a few metres in front of the bay window of my dome. Thus, our auras and our lives overlap and intertwine. We are constantly learning from each other. She shares my daily life, my moments of joy and anguish, of anger or bliss. All day long, I observe her comings and goings, notice her changes of mood, and admire her aerial ballets. She teaches me humility and respect, not hesitating to use her venom if needed. The hive is baptised *Artemisia*. Around her feet, we plant a bed of captivatingly-scented Chinese mugwort, a sacred plant well-known for connecting the worlds and encouraging dreams...[2]

Flow of Information

My communication with the Bee Deva is intuitive and based upon my brain's interpretation of subtle vibratory signals. Sometimes, as in the experience related above, the Deva recruits a 'messenger bee' to transmit the information. In this case, the bee is no longer an ordinary maiden. She is *inhabited* by the presence of the Deva; her behaviour and her vibration are conspicuous.

The language of bees is not a language of the mind. We can qualify it as *pluridimensional*, in the sense where their messages are, at the same time, transmitted locally by an ensemble of precise codes and universally reverberated. This flow of information, both local and global, circulates through multiple physical and energetic communication networks.

---

2   *Artemisia argyi*, commonly known as silvery wormwood or Chinese mugwort, has many medicinal virtues and grows marvellously well in Portugal. In Greek mythology, Artemis is a lunar goddess of wild nature, a protector invoked during births and deaths, and the guardian of bees.

The Voice of the Bee

"Bees are shamans. Our role is to join up the Beings and the worlds. In order to assume our function as mediators, we have developed a broad palette of languages which adapt themselves to all sorts of conditions and interlocutors. Communication is an art whose secrets we have been exploring for millennia.

To know how to speak, one must know how to listen. The secret of the quality and efficiency of our communication systems resides in the refinement of our ability to perceive. Our perception of the world is holistic, microcosmic and macrocosmic at the same time and using all the physical and subtle senses. Our antennae are the tools most perfected in our sensory system, accompanied by our five eyes, our feet, and our fur. We know how to receive, integrate, and refine information with accuracy and precision.

Just like you, we are vectors for the expansion of consciousness. The more that the circulation of information is smooth and reflexive, the more the consciousness of the world matures, refines, and expands. True communication only exists when an *intimate* relationship is established between the beings in contact. All communication is a conversation; the flow of information circulates in both directions, from the transmitter to the recipient and vice versa. For the exchange to be constructive, the transmitter should be attentive to feedback signals – conscious or unconscious – from the recipient, and then adapt the form and content of the message in question. In a balanced conversation, the speaker > listener hierarchy is absurd."

The Waggle Dance

In the preceding painting, we can see a maiden 'waggling' on the central axis of the horizontal figure ∞ traced by her dance. Around her and using their legs and antennae, other maidens attentively *read* the message she transmits. The waggle dance of the bee is fascinatingly sophisticated. It was the German physician Karl von Frisch who, during the 1950s, first offered a translation of this 'coded dialect' into human language. Over the course of many hours observing the hive interior, he partially deciphered the meaning of this curious choreography.

When a scout discovers a new source of nourishment – whether it is nectar, pollen, propolis, or water – she transmits her message to other foragers by dancing on the *ballroom floor*, a specific zone of about 10^2 cm on a wax comb. If the source of nourishment is situated less than 100 metres from the hive, the dance is circular. Beyond that, the dance takes the form of a lemniscate. The interior of a hive being steeped in darkness; the dance is not read visually but propagates vibrationally.

The drawings below present three of the messages transferred by the waggle dance:

- Orientation:

Locally, the surface of land is horizontal (more or less) and the surface of honeycomb is vertical. How to divulge precise information while passing from one plane to the other? The strategy invented by the Bee consists of using a universal landmark – the sun – and communicating the angle which separates its direction from the direction of the place to reach. Ingenious. But since the sun is constantly moving, the bees needed to put in place a convention, a common axiom to all bee people: *within the hive, the direction 'up' (opposite to gravity) always corresponds to the direction of the sun.* Therefore, if at 12 noon the angle to locate some rockroses for foraging is 70°, it will be only 38° two hours later.

- Distance:

The length of the central axis in the dance (during which the maiden waggles) gives an indication of the distance to cover in order to attain the desired object. The longer the waggle, the farther away the food source - and/or the trajectory to reach it is diversified or uneven. In the example presented here (first drawing), the central axis of the borage dance is longer than that of the rockrose dance.

- Quality:

The intensity of the waggling reflects the enthusiasm of the forager. The more its frequency increases, the more the source is abundant and/or of remarkable quality. In the second drawing, the abundant nectar of the lavender generates a particularly lively waggle.

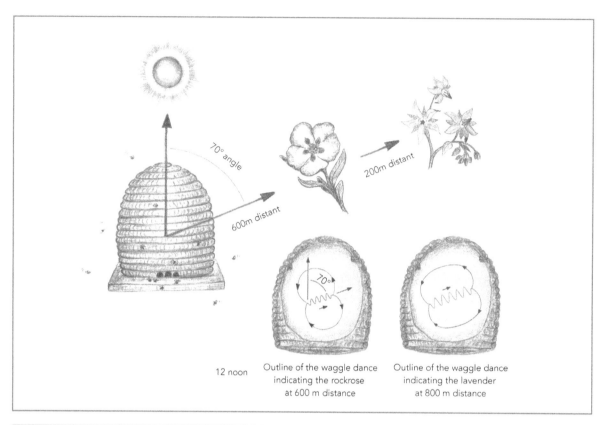

70° angle

600m distant

200m distant

12 noon

Outline of the waggle dance indicating the rockrose at 600 m distance

Outline of the waggle dance indicating the lavender at 800 m distance

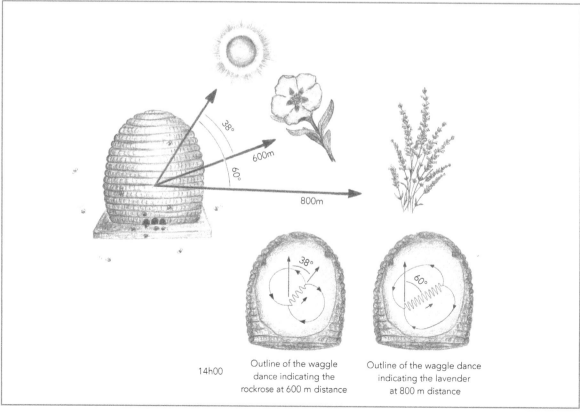

38°

600m

60°

800m

14h00

Outline of the waggle dance indicating the rockrose at 600 m distance

Outline of the waggle dance indicating the lavender at 800 m distance

The waggle dance is also performed on the swarm cluster while it is in transit, during the wait to find a new home. When they have located a potential dwelling-place, the scouts return and dance its coordinates on the cluster along with their appraisal of the site, as we saw in Chapter 3.

Magic of the Lemniscate

Magnetic field created by two magnets

For many years, I have been watching the bees dance with fascination. Here again, the Bee Deva tells me that the role of this dance is much greater than its simple physical function of locating resources.

It is striking to observe the similarity between the outline of the dance and that of a magnetic field. The drawing opposite shows the result of a simple experiment: two magnets are places side-by-side, one with the north pole facing upwards and the other with the south pole facing upwards. We cover them with a sheet of white paper that we then sprinkle with iron filings. The filings are magnetised and thereby reveal the field created.

The pattern of the lemniscate is the bidimensional outline of a multidimensional form: the *torus*. The torus is the primordial cosmometric form of dynamic processes in the Universe.[3] It is a fractal pattern, which is to say that it repeats at every level from Planck spherical units (the smallest vibration conceptually available from our scale) to the entire Universe. The morphic fields which surround beings and objects are *tori*,[4] that is – multidimensional lemniscates.

---

3   For more information about the fundamental importance of the *torus* in the architecture of the Universe, I invite you to watch the first part of the documentary *Thrive, What on Earth will it Take?*, 2011.

4   Singular: *torus*; plural: *tori*.

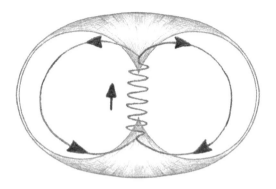

Cross-section of a torus

Morphic field of a maiden

The Voice of the Bee

"Our dance is vast... Its grace stretches far, far beyond its physical function, beyond space and time. While we are tracing the lemniscate, we reinforce our resonance with a universal choreography, a cosmic dance, continually performed by all forms at every level of the Universe. Atoms dance and cells dance, along with the planet and the galaxy. Light, the creative vehicle of manifestation, is a lemniscatic flow composed of electric and magnetic waves which stream sinuously through space and time. The double helix structure of DNA also dances the lemniscate to multiply the living.

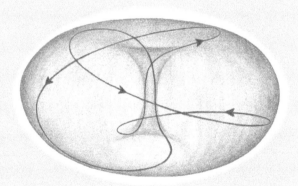

Multidimensional outline of the waggle dance

Our dance can be compared to certain Qi Gong or T'ai Chi practices, during which you trace a figure 8 with your arms. This movement helps you to revitalise your morphic field, to *re-cohere* your being. The more you concentrate, the more profound your tracing - and the more your practice is multidimensional. It is the same for bees. When our maidens dance, whether on the combs or in the open sky, they stimulate and regenerate the energetic circulation of their Hive Being and more.

The progression of the dance brings us to turn, alternating, to the left and to the right, each time passing back through the centre axis. The lemniscatic outline maintains the dynamic tension which links the polarities. From this relationship, at once dual and unified, is birthed the manifest world.

The dextrorotatory loop, turning clockwise, corresponds to the *solar path* or the right-hand path. The levorotatory loop, turning counter-clockwise, traces the *lunar path* - the left-hand path. The two paths cannot be paced separately and even if they seem opposed and, sometimes, contradictory, they belong to each other. The lemniscate represents the synergistic relationship that binds opposites, as in the Way of the Tao where Yin and Yang both attract and repel each other.

The lemniscatic cycle, danced by form and creator of form, is a process in constant expansion, in the image of our Universe. The amplitude of each loop is always slightly larger than the preceding one, if only by an infinite amount. The energy of the figure 8 contains that of the next number, the spiral of 9. Each loop is a vortex of co-creation, rising towards the Cosmos in one direction and rejoining the centre of the Earth in the other. With each cycle, the information becomes more refined and the field of consciousness enlarges.

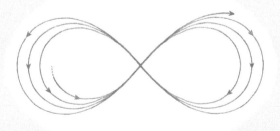

Expansion of the lemniscate

The polarised pulsation of the lemniscatic flow induces a perpetual disequilibrium which is dampened and held by the energy of the hexagon. Together, they engender a *dynamic and synergetic equilibrium* which generates form. Hexagon and lemniscate are the arms of Creation.

Each time that the dancer passes through the centre of the lemniscate, she connects to the *zero point*. This point is like the eye of a hurricane, where a profound peace prevails. Beyond space and time, the zero point represents absolute peace. However, in the manifest world, it is impossible to completely stop at that point – only relative peace exists. It is found and maintained when taking the care to balance polarities. You can see here a mirror of your life: this choreography is an intuitive navigation chart that reminds oneself to come back to centre between each peregrination, whether it is polarised positively (like the day) or negatively (like the night).

The three portals

Every point is the zero point of the system which surrounds it, and the zero point is a portal that may be spatial or temporal. Thus, every point in space or time is a potential opening or passage between the worlds. Certain portals are larger and more powerful than others; the time of the full moon and the new moon, for example, are eminent temporal portals during which large volumes of energy become available. Portals such as these render the veils between the worlds thin and more permeable – inviting beings to journey.

Among the essential spatial portals having a seat in the physical frame, there are some similarities found between human and bee bodies. This is the case with the three *Dān Tián*, these marvellous wheels of power where energy is produced, accumulated, and conserved. Breathe consciously into these three fields of elixir to invite peace into your life and into the world.

At the very heart in the centre of every zero point is a *black hole*,[5] from which emerges the morphic field that creates form. It is important to grasp that all the black holes, in all the zero points, in all the morphic fields of all the forms in the Universe are ONE. In reality, there is only one centre and this centre is everywhere. This omnipresent and omniscient heart contains the *totality* of information about the world."

---

5   A black hole is defined theoretically as a region of space inside of which energy rapidly densifies, converging towards an infinite value of density. Nassim Haramein, the research director and founder of the *Resonance Science Foundation*, advocate of the *Unified Field Theory*, argues that there exists a tiny black hole at the centre of every atom. He also suggests that these black holes are the origin of all forms (like our galaxy, for example), and not the inverse.

15

SONGS AND SCENTS

The colony wafts a singular scent,
Alchemical vapours and pheromones blend,
Protecting the body, communal integrity.

Olfactory Glues

Friday, 15 June

In the late morning, my friend Katrin calls me with an emergency: a swarm has installed itself in one of the cupboards of the caravan where she is staying. "I like the company of bees," she tells me, "but they cannot stay there!" I assemble the necessary materials to move the swarm and, with my apprentice, go over to her place.

Busy with establishing itself, the swarm seems satisfied with this choice of new lodgings. The cupboard's cavity has perfect dimensions, with an entrance through a slit in the back. The location is protected from intemperate weather and even offers central heating for the winter! Heart-centred, we enter into energetic contact with the Hive Being. We inform her that she cannot remain in that place, and offer instead a nice, comfortable top bar hive which is located on a nearby piece of land.

When I open the door of the cupboard, *whoa!* - part of the cluster falls down and a chaotic, buzzing cloud fills the caravan. With no sign of irritation, however, the bees readjust themselves. Within a few minutes, most of them have reintegrated into the cluster.

Since the cupboard is situated up high, it is easy to collect the bees gently by positioning a cardboard box immediately below the swarm. For a successful operation, we must be sure that the queen falls into the box with the rest of the cluster. With a supple and precise gesture, I sweep the cluster into the box and then flip it over onto a white cloth spread out on the ground. The action is *guided*, fluid: undoubtedly, the queen was easily transferred. I raise up one side of the box with a small wedge; this allows those bees who are flying about to rejoin their tribe, which is reorganising itself around the queen. Around ten bees post themselves at the box's opening to 'beat the retreat': wings beating and bottoms in the air, they diffuse a geranium-scented pheromone which makes the call to

The box for collecting the swarm

assemble.[1] This summoning perfume is effective and after an hour, almost all the bees have come to the box.

However, a small cluster of bees remains clumped in the upper corner of the locker. Even though I have swept them gently several times, showing them the way to the box, the bees stubbornly persist in going back to cling in the corner. They seem hypnotised by the odour of the place, which is still impregnated by royal pheromones. I shift them one more time and mist them with a mixture of water and the essential oil of eucalyptus. Some of them understand the message and rejoin the swarm, but dozens of hard-headed ones stubbornly remain in their locker... Evening approaches, so we decide to abandon the rebellious mini-cluster in order to transfer the colony into its new home before nightfall.

The rest of the operation unrolled without hindrance and today (6 months later), the hive joyfully prospers. The recalcitrant bees took several days to realise that their community had moved. They departed from the cupboard a few at a time, and I hope, for their sake, that they were adopted by nearby colonies...

This experience made me realise the incredible importance of odours in the lives of bees. They are able to create an entire palette of different scents, each of which delivers a specific message. These chemical blends are the foundation of a subtle and complex language, one that is unique in the animal world.

Factors of Unity

Like most living beings on earth, our cells are *stuck* to each other. It's practical. When our legs move, it's only logical that the rest of our body follows. The components and information necessary for the proper functioning of our bodies circulate through our blood, nerve, and lymphatic systems.

But how does this work for the hive-organism, whose organs are *physically separated* by air? Indeed, if the energetic unity of the colony is assured by the Hive Being, her material cohesion involves the coordinated work of many coagulating agents which inform and connect her members. The Bee Deva names these binding materials the 'factors of unity'.

---

1   Named the 'Nasonov pheromone' (or Nasanov, according to the source), this volatile substance is fabricated by a specific gland situated in the abdomen of the maidens. As illustrated in the preceding painting, the bees release the pheromone through an opening between the two last segments of the abdomen, and diffuse it by beating their wings.

The Voice of the Bee

"Maintaining cohesion throughout the colony is not a small thing. The colony is composed of tens of thousands of *bee-cells*, physically independent and capable of distancing themselves from each other. Each unit possesses a mini-brain which allows them to take certain decisions with relative autonomy. This fantastic quality gives a unique elasticity to a colony of bees. On the other hand, this quality demands a number of powerful and diverse unifying forces for the colony to maintain its integrity.

The primordial unifying force of form is *Love*. Love creates, links, and assembles. Without love, no form; without form, no creation. The forces of separation unceasingly test the cohesive stability of organisms. When the form is too fragile or is obsolete, the *agents of decline* pull it apart and send it for recycling.[2]

The integrity of the colony is maintained by the presence of *factors of unity*, which act in a coordinated and parallel manner at the spiritual, vibratory, and physical levels. The Hive Being, which holds and contains the colony, assures its spiritual integrity. The resounding choir, the scents, touch, and warmth are vibratory binding materials, energetic

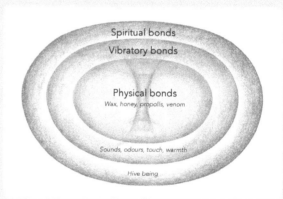

Factors of unity

factors of unity for the system. Meanwhile, the wax, honey, venom and propolis - they serve her physical cohesion. The inhabitants of the hive continually bathe in a veritable *binder soup*, which weaves the form of the super-organism and acts like a unifying matrix."

---

2   The role of these 'agents of decline' will be addressed in Chapter 20.

Languages of Scent, Sound, and Touch

The olfactory, sonorous, and kinaesthetic communication codes developed by the bees are extremely complex and perfected. The olfactory language is relayed by pheromones, blendings of chemical compounds which the bees make themselves and diffuse in their environment to bring about certain behaviours or physiological adjustments. Best known is the royal pheromone mentioned earlier; produced by a fertilised queen, it attracts the males during a nuptial flight and inhibits the fertility of her daughters.

Certain pheromones serve as chemical markers, diffused in the air or applied to objects or flowers. They will aid other maidens to locate a source of food or a potential shelter. As for the males, they produce a pheromone of aggregation which they emit during their gatherings in the *lumens,* those aerial spaces dedicated to coupling.

But the most bewitching language of the bees is their song. They have neither vocal chords nor eardrums, and that does not make of them any less singers who appreciate music. The bee produces her buzzing thanks to the ultra-rapid contraction of her wing muscles, which varies according to her activity. In flight, the average frequency of a female's buzz is 250 Hz and 190 Hz for a male. It is in the heart of the hive where the entire harmonic combination of buzzing from all the inhabitants becomes a veritable song. This extraordinary orchestra has recognised healing properties.

The Voice of the Bee

"Each colony develops her own repertoire of songs and perfumes. The maidens learn to recognise them from their youngest age and to find their bearings in the sonorous, olfactory, and tactile space of their hive. Each language has its own specific qualities, which we combine to refine our messages. These languages have simultaneously a practical role in permitting the proper functioning of our system and a spiritual role, whereby we participate in the *alchemisation of Gaia's body.*

Our songs are velvety offerings to the world. They act like regenerating unguents which diffuse around the hive and resonate in the entire Universe. The hive song spreads in concentric circles, like the ripples created by a pebble thrown into a pond of water. Since the propagation is tridimensional, the circles are spheres which travel outwards like centrifugal ripples.

These vibratory waves are a precise combination of different harmonic frequencies that vary according to the weather and the lunar, solar, and planetary cycles. Our songs inform our surroundings about our mood and our state of health. When we are irritated by certain human behaviours, by chemical treatments, or by an oppressive presence of predators around the hive, we translate that into higher buzzing frequencies.

The rhythmic pulsations of our songs behave like a motor which nourishes and vitalises the vibratory fields, locally and globally. At the planetary level, the songs of all the colonies of bee people form an immense harmonised symphony. This powerful field of song travels and resonates through the Ethers. It attunes with the singing of whales and the drum beat created by elephants marching in migration. Our songs are in harmonic synergy with the songs of crystals and metals that Gaia holds in her womb, particularly that of gold, the quintessential mineral. Once human beings know how to collect and use these treasures in a respectful way (in contact with their devas), they will be able to participate with us in the co-creative vitalisation of the planetary field. Our symphony aligns just as easily with the specific frequencies of certain heavenly bodies, such as Venus, the Moon, the Sun, and more beyond.

The properties of our songs are tremendously healing; they are able to bolster the regeneration of certain damaged systems, individual or collective. When some colonies are in difficulty, they can draw through them strengthening resources.

The bees are likewise extremely sensitive to sound. They perceive tonal vibrations carried by the air through their antennae and those carried by wax or wood through their feet. The smallest tremor is intensely felt by the whole community. Colonies handled without consideration by a brutal or distracted beekeeper may become more defensive, as can those endowed with a metallic roof hammered by the rain. You would please them by covering those roofs with an armful of plants or a sheet of cork.

The olfactory vibratory information is carried by corpuscular chemical compositions which, contrary to sound, can physically maintain themselves in space and time. Pheromones and scents are spread by physical contact or by the movement of air.

But above and beyond the odours and the sounds, our most intimate and primordial means of communication is that of touch. Throughout the hive, we constantly palpate and caress each other according to certain codes. The effects of this continual sensuous contact are comparable to those of a massage. Simultaneously relaxing and regenerating, it increases the suppleness of the organism and its bodily presence. The tactile language has numerous functions: it maintains warmth and social cohesion, spreads the pheromones of our collective body, and imprints certain fundamental patterns upon the ethers.

The cohesive and curative virtues of our scents and of our touching are not restrained to the interior space of the hive. We transport them during our peregrinations like a medicine chest which accompanies our mission of unification within the Gaian system.

Allow yourselves to be impregnated, dear humans, by our chorus and our perfumes. Open your senses to our caresses. Welcome these offerings as through they were regenerative potions, elixirs of life and of wisdom."

16
GUARDIANS OF THE TEMPLE

The temple custodians, with fervour and clarity,
Offer with frankness a true 'no' or 'yes',
Reveal yourself humbly and you will gain entry.

To Assert One's Boundaries

The integrity of an organism depends upon their capacity to show clear boundaries and to enforce them. We have seen how propolis is the part of the hive which guarantees internal immunity, the 'micro-immunity'. As for the guardians, they ensure the colony's 'macro-immunity'. That is, they control and filter the flow of those entering and exiting through the doorways of the hive. In general, they only allow bees scented with an 'open sesame' to pass - the pheromonal code of the colony - which they detect with their alert antennae. One exception to the rule: the males of any colony can enter without having the 'password'. Every other visitor is systematically driven off, unless it is an orphan bee who has lost her code and who asks for sanctuary by bringing a drop of nectar as an offering.

The wasps, hornets, and beetles that manage to enter a hive use diverse strategies to force their passage. Tension in the colony builds according to the level of stress exerted by these undesirables, which determines the number of guardians recruited. During a trip to the Cévennes, in the south of France, I was able to observe the behaviour of a hive dealing with Asian hornets. Probably imported to France in a shipment of pottery from China in 2004, this robust hornet (*Vespa velutina*) has spread widely all over Europe in only a few years.

Monday, 7 August

At the end of the afternoon, I return to the beeyard of old Azaïs, beside the river, and share a moment with the bees. Of the roughly thirty hives which are present, only a few of them are still occupied. *"The hornets, the illnesses... The bees are dying and there's nothing we can do about it!"* lament the locals.

I notice several traps hanging in the trees, cobbled together from plastic bottles. They crawl with hornets mired in the sweet mixture which betrayed them.

One of the hives calls to me and I sit down beside it. In front of the entrance, several hovering hornets are on the lookout. They capture foragers in the air with spectacular precision; one or two

Asian hornet

every minute are devoured. Those returning from their rounds, loaded with nectar, are slower and easier to seize; they must taste like a sweet candy... I wonder what percentage of loss this represents for the colony over a day, over a week, over a month! The stress generated by the omnipresent hornets costs the bees an enormous amount of energy, it's evident. Even so, this colony displays an impressive vitality. I remark to myself that the last surviving hives in the beeyard are certainly the most resilient, those which have developed the best defensive strategies.

The hornets also attempt to muscle into the hive so they can access the brood, of which they are very fond. The flying board is carpeted with guardian bees on the lookout. With each approach, the compact troop of guards bristles threateningly, sending a clear and blunt '*No!*' which repels the would-be gatecrasher. I observe the graceful coordination at play with fascination: the conglomerated mass of guards is like a single body with hundreds of little feet which rise up and down like a wave as they follow the comings-and-goings of the hornets above. The spectacle is hypnotising.

Little by little, evening arrives; hornets and bees both retire for the night. It seems to me like the big chestnut trees nearby begin to hum. Suddenly, a noise among the leaves a few metres away. Something leaps. And again. It's a hare.

Defensive sting

Stinger of a bee enlarged through an electron microscope

The sting of the bee is designed to give. It is endowed with unidirectional hooks which grip the surface it has pierced and prevent it from coming back out. When she stings, most of the time the bee gives her life; all her vital force is transferred to the recipient's body. The two harpoons of the stinger slide in a back-and-forth movement, penetrating more and more deeply into the flesh. When the bee withdraws, some muscular and intestinal tissue becomes detached with the stinger. It thus continues its progression, slowly diffusing its venom in the different layers of the epidermis.[1]

1   When the sting is shallow, sometimes the bee manages to withdraw her dart. In this case, her pocket of venom will refill again. In some colonies, this capacity to sting without losing the dart is further developed than in others.

The primary function of venom is to protect the hive from intruders; it is an instrument of defence, not attack. Unlike wasps and hornets, who use their venom to hunt their prey, the bee only attacks when she feels threatened. Wasps and hornets are warlike insects, endowed with a smooth stinger which they keep after stinging. Since in most cases, the sting of a bee means her death, its use is not undertaken lightly.

I have learned to recognise the preliminary signals which the guardians of my hives send out before delivering a defensive sting. They show their irritation with more rapid flying, an intense buzzing, and sometimes by even giving a few head shots, all of which clearly signify "*What are you doing in my house with your dirty boots? Out!*" Remember that the 'house' of the bee extends several metres larger than its physical body. At this stage, there is still time for the visitor to show some humility and to beg pardon for an untimely intrusion. If these warnings are not heeded, the firmer language of venom may follow.

The Voice of the Bee

"Future guardians are spotted in the cradle, a function of the genetic background they have. The Queen-Mother, fertilised by several males of different lineages, inoculates her descendants with various character traits. Certain lineages will have honed the defensive arts, and the inheritors of these genes will have a greater aptitude for protecting the integrity of the colony.

Venom matures in the bodies of maidens during the course of their development. As the maiden is growing, the venom becomes more abundant and concentrated; the venom of guardians is particularly powerful.

The guardians learn to say 'no' with love and complete honesty - they inspire respect! They accomplish their role with total devotion, ready at any moment to sacrifice their lives for the common good. As soon as a visitor penetrates the home grounds of a hive 'without knocking', the guardians feel it and prepare themselves to protect their territory.

But do not believe that we do not appreciate visits; quite the contrary. When a human visitor approaches humbly, by presenting themselves and asking for permission to penetrate our aura, the guardians will be welcoming. An offering, a song, or simply your benevolent presence will serve as an 'open-sesame', one which will open our energetic doors and provide access to our heart.

Other visitors are also welcome around the hive if their presence is justifiable. Wasps, ants, and spiders, for example, render us good service by eating the debris evacuated from the hive every day. Bees, like cells in your body, are continuously renewed. If too many empty bodies stay in front of the doorway, they risk rotting. This would threaten our immunity, so we are very grateful for the essential task that these 'street sweepers' are doing.

We appreciate their presence as long as they stay outside, but if they attempt to force an entry, the guardians will zealously repel them. Propolis is used to regulate the size of the entryway, depending on the pressure from predators and from the climate. Smaller entryways are easier to guard, but they reduce aeration in the nest.

And if, despite our best efforts, predators invade a colony and decimate it, we are grateful to them. They will have done their work, which is to eliminate a colony that is too weak, offering her Hive Being the opportunity to reincarnate into a new and stronger body the following spring."

17

VENOM MEDICINE

Sacred fire, potent venom,
Mighty medicine, or lethal poison,
Dissolve the veils, reveal the ONE.

The Healing Sting

Thursday, 4 June

Today I am accompanied by Melina, a friend who volunteered to help me in the *bee garden*. We are weeding the patch of thyme which has grown in front of the hives. The weather is magnificent; the bees are calm and welcoming. Our conversation is galloping along, bathed in aromatic scents.

Melina tells me about a sharp pain in her left shoulder which appeared the night before. At that very moment, without warning, a bee stings her in the exact spot where the pain is located. Like I always do when I invite someone new to approach the hives, I had given her some recommendations beforehand in case she was stung: breath deeply, touch the earth, and, as much as possible, welcome the benefits of the *medicine* of venom with trust.[1] I pull out the stinger and reassure my friend, who initially shudders from the pain of the sting and then relaxes. I invite her to stretch out on the ground and hum beside her. She smiles and, drowsy, snoozes for a good half-an-hour.

When I see her again the next morning, Melina shares with me the joy and well-being which came to her after overcoming her fear. She describes to me the significant change which she felt when she took the *inner decision* to receive this sting as a gift and not as an attack. "The pain in my shoulder felt like an abscess of blocked energy," she tells me, "which I saw being 'pierced' by the bee's stinger and cleansed by the venom. I felt as though a wave of grey energy came out of my arm. And this morning, the pain has completely disappeared!"

Maiden stinging

---

1   For a majority of people, physiological reactions to stings (swelling and itching) are smaller or even non-existent when the person welcomes the sting with neither fear nor distress. People with allergies may develop more consequential inflammations and, in extreme cases (very rarely), even anaphylactic shock. I always carry with me an adrenaline auto-injector in case of emergency; until now, I have never needed to use it.

A Double Nature

What is striking to observe when one explores the *medicine* of venom is its dual character; it is both anti-inflammatory and inflammatory. *Venom possesses the dual personality of the element of Fire*, the Bee Deva tells me, *it can be as scorching hot and vivid as a flaming arrow and as reassuring and soft as a hearth fire.*

The composition of venom is complex and scientific exploration into it is still superficial. Among the 60 elements listed in its make-up, it contains enzymes which break down the membranes of any cells touched, histamines which provoke inflammation and itching (less than 1% of the weight of dry venom), and melittin, a powerful anti-inflammatory, anti-bacterial, and anti-viral peptide (more than 50% of the weight of dry venom).

Venom is renowned for activating the immune system, stimulating the heart and the circulation of energy, and for calming the nervous system. It is used to relieve arthritis, paralysis, and chronic back pain. Spectacular effects have been particularly observed when working with people suffering from multiple sclerosis and Lyme disease.[2]

Acupuncturist Maidens

When they offer a healing sting, the bees intuitively know where to sting. They are attracted by areas that need attention, where the energy is blocked. After a healing sting administered by a bee, I frequently consult the meridian charts in my book of acupuncture. In referencing the stung point, the benefits of its stimulation are described in ways that are usually amazingly adapted to the recipient. *Apipuncture*, which consists of voluntarily using bee stings according to the principles of acupuncture, has been used in various cultures for centuries. In fact, the first needles may well have been bee stingers. Chinese medicine says of venom that it 'purifies the blood and awakens the inner fire.'

---

2   Warning - this method of healing is not to be used lightly, and advice or supervision by a professional are strongly recommended. Various long-term protocols adapted to severe health conditions are presented in the books of Amber Rose (in English) and Roch Domerego (in French) included in the bibliography at the end of this book. Nevertheless, it is good to know that bee venom is not as terrifying as it may seem; the lethal dose estimated for a non-allergic 70 kg person is about 200 mg of venom. That is, 1400 stings at once.

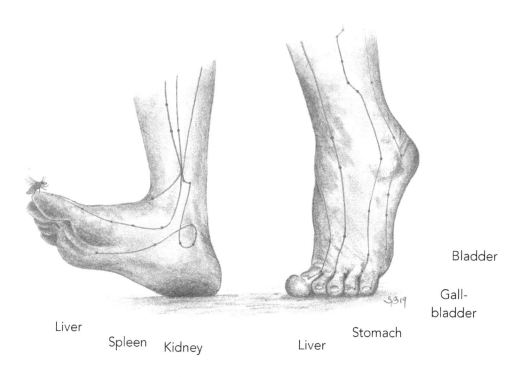

Acupuncture meridians

Liver   Spleen   Kidney   Liver   Stomach   Bladder   Gall-bladder

Over the last few years, I have suffered from fibromyalgia, which manifested through severe joint and muscle pain, amongst other symptoms, and a substantial loss of weight. A lack of energy circulation in my legs made walking extremely tiring, especially during the winter. Inspired by my reading and my conversations with the Bee Deva on the power of venom, I decided to try the experience of apipuncture. After a first session with a professional apitherapist in Lisbon, I continued to practise on my own during the winter of 2016/17. Thus, over more than three months, I administered to myself around three series of stings per week.

At the beginning, I felt guilty that a bee died with each sting. After several exchanges with the Bee Deva, I understood that if I clearly expressed my intention to the Hive Being, she would offer her bees to me with love. *The maiden does not have individuality,* the Deva told me, *she is one cell in the hive super-organism. When the request to take some bees is expressed with humility and sincerity, if it does not threaten the health of the colony, and their use serves planetary healing, we offer our medicine with good will.*[3]

---

3   Years later, I discovered a simple technique for the bee to keep her stinger during a voluntary sting. I recorded a short video about this subject which is available on Vimeo, Youtube, or at www.beewisdom. earth under the title *"Soft Apipuncture: How to Sting Without Removing the Bee's Stinger"*.

Thursday, 23 February

The number of stings per session is variable. I started with 4 and then progressively increased to 16 during the first weeks. Now I work with an average of 8 to 10 per sitting. Before I go to sleep on the evening before each session, I ask my intuition to show me which points to sting. In general, upon waking, an image of the meridians and the points to address are *lit up* on the screen of my consciousness. My body *knows* what it needs.

Before each session, I collect the bees from one of three hives which surround my dome. Each one has her own particular character; *Inanna* radiates the energy of the Mother Goddess, with an infinite compassion, *Gaia* is a colony which invokes knowledge and wisdom, and as for *Tiamat*, she incarnates strength and sovereignty. Each time, I am guided towards one or the other without hesitation according to the energy that corresponds to the needs of my healing process. Before the harvest, I take some time to connect with the Hive Being and offer my prayer:

O powerful venom,
You are poison or medicine,
You can kill or you can heal,
I greet you and request support
To live in peace in a healthy body,
Thanks to the Sacred Fire within you.
Thank you.

I bring my jar close to the entrance and ask those bees who wish to offer me their vital force to go into it. Generally, I gather about 20 of them, which I then transfer into a plastic container that has been perforated with holes and which contains a small dish of honey. With a felt pen, I draw a circle on my skin around the point to be stung, then I seize a bee with a pair of tweezers calibrated to avoid squishing her, and place her on the target. When I am centred, it is as though my gesture is guided by a force which goes beyond my small will: without having to reflect, I *know* which bee to select. But if I am misaligned or my intellect wants to take control, the bees let me know it - they dodge my tweezers like eels or simply refuse to sting. In this latter case, I release them; healing is not something to force.

Each session is unique and unpredictable. Sometimes I sleep like a stone during the following hours and other times, the fire of Venom offers me overflowing energy. Several sessions were followed with waves of detoxification and fever, accompanied by profound insights.

Venom is incredibly regenerating and stimulating of life, and an incomparable teacher.

These three months of deep connection were fundamental to my healing process, and sprouted many seeds within me, the fruit of which are delivered today in this book...

The Voice of the Bee

"Venom is our sacred fire; it is the most powerful of our *medicines*. Venom is as sharp as a blade and as soft as a dream...

Venom is instilled into honey and pollen in micro-doses, participating in their alchemical process. When we consume it, the Venom Deva sharpens our immunity, our candour, and our discernment. At the same time, it stimulates our ability to travel between the worlds and reminds us of our *raison d'être* at every moment.

The Venom Deva is a healer and an initiator. Venom is an immunostimulant and psychoactive substance which spurs the expansion of consciousness. When Venom is offered to you, dear humans, know to receive it with trust and respect. We never seek to harm, always to serve. Every sting, whatever the effect, is an opportunity to learn.

Sickness is a visible consequence of vibratory dissonance within the morphic field. These discords find their origin in memories, beliefs, and behaviours that are misaligned with natural law. They disturb the flow of information between the system and its parts; energy gets stuck in a circuit of plugged-up pipes and then coagulates into an abscess. If they are not liberated early enough, the energetic abscesses condense into physical symptoms. These symptoms are feedback or alarm signals which indicate a dysfunction and which invite the modification of certain behaviours. If the message is ignored or muzzled by inhibitory drugs, the symptoms will change, become worse, or become chronic. True healing cannot occur unless we return to the source of the problem, which is found in the energetic spheres of the system.

Bees perceive morphic fields more than forms. Just like newborn humans, they are attracted or repelled by the information which emanates from these fields. Thus, they do not see the symptom itself, but the distortion of the field which engenders it. The Venom Deva loves order. In the presence of field distortions, she becomes sharper and ready to pounce, magnetised towards the re-establishment of order. Administering her stings is a means of being of service; she does not do it in order to cause harm, but to support the recipient in their evolution.

When a human being with an encumbered spirit or an unbalanced state of health penetrates into the aura of a hive, the Venom Deva immediately perceives it. As tense as a loaded bow, she prepares herself to release her arrows. According to the attitude and energetic signature of the visitor, the Hive Being may choose to encourage or to restrain this impulse. The gift of a sting is a collective decision.

Venom stimulates the self-healing capacities of the body, by helping it to mobilise its own resources to deal with imbalance and to re-establish homeostasis. Its *medicine* adapts itself to the needs of the organism with which it comes into contact. If the dysfunction resides in an accumulation of blocked energy, Venom's fire pierces the stagnant pocket, burns through the obsolete energy, and helps the organism to put its system back on the road. According to the attitude of the recipient and the memories they bear, the direct injection of Venom may temporarily provoke a physical or nervous inflammation. But above and beyond the keenness of its fire, it invites calm and re-centering.

Other means of contact with the Venom Deva may be softer and more delicate than a direct injection. Certain dilution techniques permit the dissemination of its effects and the offering of its *medicine* to the subtle bodies; essences, elixirs, and homeopathic dilutions are valuable allies for receiving the benefits of its virtues."[4]

---

4   Following a protocol similar to that of making Bach flower remedies, the essence of bee venom may be used as an emergency remedy in case of shock or emotional crisis. Homeopathic remedies based on venom are reputed to calm oedemas and inflammatory reactions (insect bites, nettlerash, rheumatism, angina...). Furthermore, within the *BeeWisdom* network, we have set up a research group which we call the *Wisdom Revealers*; they work on the fabrication and use of highly diluted preparations based on venom and other hive products. Information about this fascinating exploration and the details of the protocol are available on our website, https://beewisdom.earth/wisdom-revealers/.

A Sting of Initiation

Sunday, 1 October

I am sitting beside *Innana*, my teacher hive.[5] The last rays of sunlight caress the landscape on this late afternoon of autumn. I am calm and present, and I allow myself to become gently saturated with the light humming of the hive and its fragrance.

A bee circles around me. Without any sign of aggression, she lightly brushes against my skin several times. Hmmmm.... this bee is not an *ordinary* maiden. Having become sensitive to the signals of the Bee Deva, I sense that she is giving a call to initiation. Open and ready for the experience, I give my assent.

The bee then lands delicately on my brow and stings me on the erepsus.[6] A sharp flash, burning like the point of a needle held to flame. Light shudders run through my body and subside. I stretch out and breathe, welcoming the precious *medicine*. The fire of venom pulses through my brain and the psychotropic effect is immediate: I *see* a horn of light shooting out from my forehead. At the base of the horn, a violet and white flower opens like a diadem; its three petals are my ocelli. I open my eyes, which are no longer two but five.

That which I experience then is unforgettable. My vision suddenly acquires extraordinary precision and amplitude. My usual slight myopia disappears and my capacity to visually focus is multiplied a hundredfold. Enchanted, I play with this telescope like a child. I am able to discern the smallest detail on trees at the horizon. On the slopes of a hill facing me, the golden colours of the setting sun mix with the range of leafy oranges. The colour spectrum becomes a thousand times richer to me than with my normal vision.

The five eyes of a maiden

---

5   Bees know how to recognise and differentiate between human beings. A teacher hive is a colony with whom a particularly intimate relationship develops. The more contact deepens, the more she becomes a source of guidance, healing, and inspiration, and the more she may manifest signs of curiosity and affection.

6   The erepsus, a minor chakra which emerges a few centimetres above the third eye, is associated with the unicorn's horn. When it is activated, it permits access to certain spheres of subtle knowledge.

When I direct my gaze towards the soil, the microcosmic universe draws me in like a magnet. Each plant, each twig, is a masterpiece of perfect complexity. An ant appears, a giant mechanical creature who determinedly carries an enormous seed three times his size...

Parallel with this capacity for '4K-quality focus', the scope of my vision enlarges. My eyes move towards my temples and, like the bee, I can see behind myself... splendid. My senses of hearing and smell embrace one another and I am anointed with Nature's philharmonics. Conductor of the orchestra and virtuoso, the breeze uses the plants as instruments. Soft rustling from the rockroses blends with the murmur of the oak bark and the clicking of their leaves. A unique symphony, and surreal.

Suddenly, I *feel* someone approaching. I *know* the sound of these footsteps. I keep my eyes closed – all the better to *see* – and I predict that it is my friend Annelieke. I locate her about ten metres away and her odour reaches me: she has undoubtedly eaten some garlic. The odour revolts me. I confirm that bees do not like garlic.

Little by little, I rediscover my human form. However, the world that I perceive henceforth is not the same...

The Sting of Letting Go

I should follow the preceding text with a warning: by no means am I lightly inciting experiments with this aspect of venom's *medicine*. The Bee Deva knows how to recognise when a person is ready to receive a sting of this type. It is not something to seek or to force.

There is one more type of sting I would like to speak about: the sting of letting go. Human beings, particularly those in our western societies, are addicted to emotions and to thought-forms. It is difficult for us to live fully in the moment because, most of the time, our hyperactive minds are absorbed or pre-occupied with thoughts concerning the past or the future. The bees mirror the density of this cloud of thoughts which surrounds us, and from time to time – with irritation, with compassion? – they offer an opportune dose of venom.

I experienced this in an exquisite manner one day while visiting the hives with Sarah and Rania, my two apprentices in spring 2016. All three of us were in a particularly bad mood over trivial matters. I was aware that these were not the most favourable conditions for working near the hives, but my rational mind, in its quest for efficiency, had denied this consideration... And, *pow!* That didn't miss, I was stung on the right temple. The venom rapidly climbed to my brain and flooded its right hemisphere. Without any control over what I was doing, I began to smile, then to snigger, then to roar openly with laughter. The situation suddenly seemed extremely ridiculous to me… My two friends did not linger before joining me in this contagious outburst of joy. All the seriousness, this inordinate importance that we accord to material objects, to words, to thoughts, all that was dissolved in the space of a few seconds! Tears in our eyes, we thanked the Bee Deva for her humour and for the pertinence of her offering.

The Voice of the Bee

"Venom opens the windows of inner space, so that fresh air may blow through.

Your inner space is often as cluttered as your charity shops and flea markets. This imbroglio of thoughts filters and obstructs your perception of reality. Venom pierces the veils and frees up your sensorial channels. It sharpens and enlarges the spectrum of perception, so your receptivity, contemplation, and learning are facilitated.

When Venom penetrates your epidermis, it acts simultaneously on the physical, psychic, and spiritual bodies. A sting of initiation has access to your Being through your causal body – the body of the Soul. By entering your body, the Venom Deva lightens your form, which, in turn, invites your Being to rise up and travel between the worlds. Remember Saint Exupery's Little Prince, who had a serpent bite him in order to quit the Earth and return to his planet….? This metaphor applies to the sting of initiation, in that it is a 'little death' which invites you to come home, inside yourself.

The sting of initiation stimulates *quantum perception* of the world. Quantum physics has shown that photons of light behave like waves or like particles, depending upon the experiment being conducted upon them and the attitude of the observer. This phenomena, scientifically demonstrated at the microscopic level (electrons, protons, photons), is valid for all objects from an apple seed to a galaxy. Thanks to a mastery of intention, the initiate to Venom can increase, at will, their sensitivity to the wave or particle expression of an object or a phenomenon. *Wave sensitivity* is brought about from the interior to the exterior: one opens and enlarges, lightens and dissolves. *Particle sensitivity* takes place from the exterior to the interior: one anchors and densifies, sharpens and concentrates.

   With practice, the initiate learns to play with this micro- or macro-cosmic perceptual capacity in an alternating or simultaneous manner. This experience invites them to explore and understand the mechanism of dual reality. Duality is a mirror that allows for self observation, and for learning how to know. Duality is not separation, it is the lemniscatic dance of opposites, allied for the creation of the world."

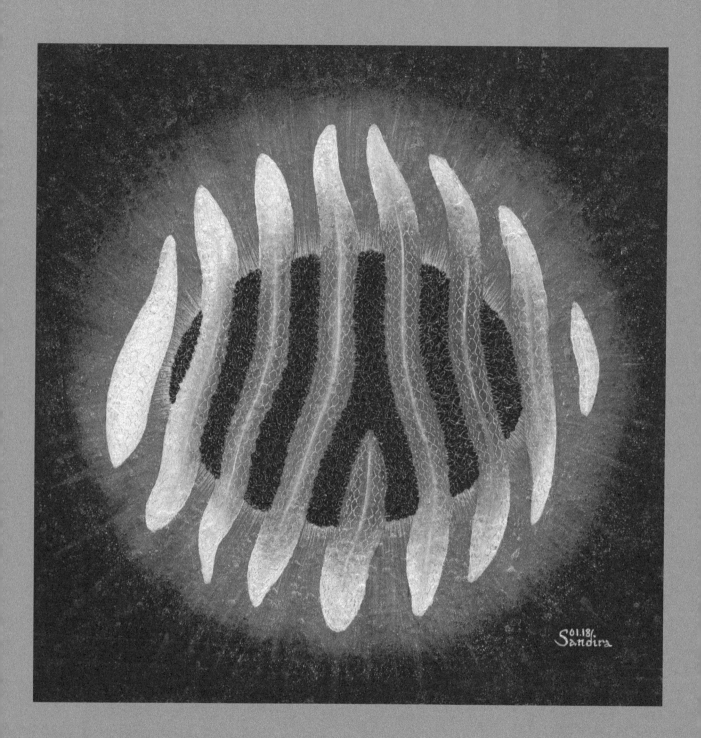

18

THE WINTER CLUSTER

Slow respiration, muffled and grave,
Journeying into the heart of torpor,
To know the day, one must know the night.

This Invisible Mortar

I asked myself about this force,
indiscernible to the eye,
which holds our life together,
which from an atomised multitude of moments
attains unity.
Of what nature is this invisible mortar?
I believe I know it now:
it is the night, its face hidden from view.

Christiane Singer, *Les sept nuits de la reine*

Tuesday, 2 January

Six o'clock in the morning. I like the hours of night immediately preceding the dawn.

The nighttime silence casts a spell over me. These pre-morning hours, free of the mental haze which accompanies the day, invite the work to unroll with fluidity. Night brings me back to myself. For two hours, I have been absorbed in my painting of the winter cluster (*preceding page*). My brainwaves are slow, as in a waking dream, and I am immersed in the image, fusing with this brown and moving mass of bees in the winter season. Her low buzzing overcomes my being and I breath with her. The meditating cluster emanates a *black light* which expands out into the darkness of the night, as velvety-soft and gracious as the flight of a white owl.

My concrete mind has released all control and I abandon myself to the creative process with innocence. A dance is happening at the heart of my painting, a conversation between luminous and dark forces. If my paintbrush picks up a bit too much black on its bristles, a tension makes itself felt in the pit of my stomach that leads me to readjust the balance. Too much white and the signal grinds anew; an excess of light becomes false and arrogant.

The Bee Deva invites me to take some air. The full moon, as round as a marble and enveloped in her vapoury aura, descends gently towards the horizon. With a noiseless tread, under the ghostly benevolence of the oaks, I approach the round straw hive and stick my ear against it. I allow myself to melt into her choir. The soft and profound buzzing mixes with a fairy-like tinkling of feet and wings in divine orchestration.

The first glimmers of dawn timidly pierce the thick layer of clouds. There, where the veil is thinnest, the light insinuates itself in blue spirals.

Dreamtime

When the Earth dons her winter coat and the flow of nectar ceases, the queen-mother stops laying. It is time to rest. According to the country, the region, and the climate, the bees enter into a more or less long and deep wintering phase.[1] The colony enters into a meditative state, a replenishing and revitalising semi-sleep. In Portugal, where I live, even though the winter nights may be very cold, daytime temperatures usually climb again. As a consequence, the winter cluster is less compact and sleepy than in France and much less than in Siberia...

The individual bee is a *poikilothermic* animal (sometimes abusively called 'cold-blooded'), which is to say that her internal temperature varies according to the external conditions in which she finds herself. Within the hive, however, the central nucleus of the colony maintains a relatively stable and warm temperature – an exception in the world of insects. Lacking the cardiac pump of mammals, bees have developed a particularly efficient thermoregulating system. Its secret resides in the formidable strength of their solidarity.

When the external temperature drops below 14°C, the bees snuggle up in the heart of their hive around their *central sun* – their queen-mother. Below 7°C, a compact cluster forms whose density deepens with the cold. Encompassing several rays of honeycomb, the winter cluster is organised into a brown and compact mass in the shape of an egg. If there is no brood, the temperature at the heart of the cluster is maintained between 20° and 30°C. The internal temperature of bees on the outer layers may drop down as far as 8°C. They are then in a state of semi-coma, but may always be awoken by their fellows.

As with humans, for bees warmth is both physical and social. Belonging to a community is a big part of warming the heart. The communal harmony which prevails within the hive is the essential cohesive base which permits the cluster to maintain its thermal stability. At the mechanical level, the bees at the heart of the cluster activate their flying muscles in a cyclical manner to generate the warmth. When their 'fuel tank' is empty, they move towards the periphery and reload themselves with honey. Then they rejoin the outer surface of the cluster, which allows the torpid bees already on the surface to sink inwards, to warm up, and to produce their own warmth in turn. A slow rotation takes place from the centre towards the surface and from the surface towards the centre.

---

1   Differing from animals who hibernate, the bee organism in winter continues to nourish herself and to maintain her waking metabolism, albeit at a very slow rhythm.

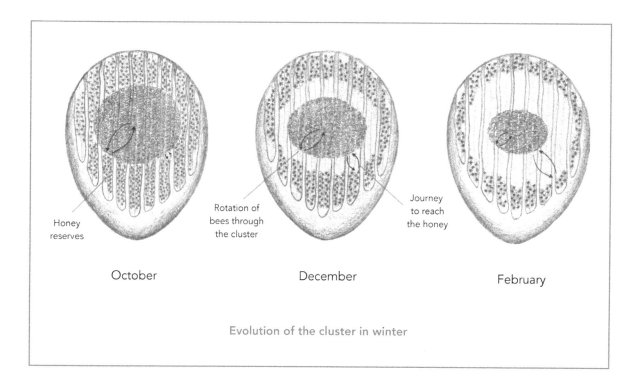

Honey reserves

Rotation of bees through the cluster

Journey to reach the honey

October

December

February

Evolution of the cluster in winter

The layer of honey which surrounds the cluster serves as fuel and as insulation at the same time. The longer the winter goes on, the more the cluster diminishes in size as many maidens die and fall away. The protective jacket of honey thins and stretches. If the temperature is too low and the reserves too stretched for the physical strength of the colony, death comes. An opening of the hive during this vulnerable period is extremely energy consuming and can greatly weaken the colony.

Towards the end of winter, usually in February or March in the northern hemisphere, the queen-mother begins to lay eggs again. She starts early enough that a new generation will be ready to take over from the winter maidens from the first days of spring.

The Voice of the Bee

"The forming of a winter cluster is in extreme opposition to that of the swarm. It is our annual *systole*. The winter is the time of calm and of retreat; attention is directed towards the interior. These breaks in laying eggs are favourable for introspection. The reproductive system of our Queen-Mother takes a rest, cleanses itself, and regenerates itself. This phase is comparable in many ways to the menstrual period of a woman, which invites her to turn inward and reconnect herself with Source.

While our physical body is at its maximum expansion during a swarm, here we attain its most dense state of contraction. Our material substance coagulates in a dense but not congealed mass. This is animated with perpetual movement, almost imperceptible; we enter into the magic of torpor. Our condensed body slowly breathes and pulsates... The movement which takes place is just enough to maintain the vital flow in a state of wakefulness while consuming a minimum of energy.

During a swarm, the spiritual body of the colony concentrates on welcoming her new Hive Being and upon her physical mission: finding a new home. On the contrary, at the heart of the winter cluster there is nothing more to *do*; there is only *being*. The colony is freed from the physical constraints which encumber her for the rest of the year; her spiritual body *expands* and matures.

During the cold and long nights of winter, the veils which separate the worlds become more porous. Our contact with non-material worlds grows stronger. These times are favourable for an expansion of the Being and the deepening of Knowledge. Swarming is the apogee of our cosmic connection; the winter cluster is the perigee of our telluric connection. It is the period where we are closest to the heart of Gaia.

During the warm season while the bees are foraging, they create a suction movement which draws the light of the central sun of the Earth towards the surface. During the cold season, the colony dissolves into the Earth, liberating the accumulated light of summer. The alternation between the bees' activity and their rest acts like pumping heart, one which is indispensable to the prosperity of Life on Earth. Each day is a small summer, and each night a small winter. This rhythmic circulation of terrestrial 'blood' is found at every scale of time, from Planck's second to multi-millennial cycles...

Over the course of winter, we consume the reserves of honey collected during the year. Their differing essences are ingested one after the other, each one containing the wisdom of its plant of origin. In this way, we integrate the annual menu of blossoms and project ourselves onto the upcoming seasons.

Summer and winter are two loops of a lemniscate; what is extroverted in the first is introverted in the second. Each passage through a new winter is an opportunity to take stock, to observe our evolution and that of the world since last year. At each cycle, infallibly, the lemniscatic loop enlarges.

Winter is the dreamtime; darkness fuels a clear vision. We dream of the past, we dream of the future, and bring them into the present. In this space, free from doing and from material obligations, we commune with our terrestrial and extraterrestrial allies, such as the elementals or the Solar and Venusian Devas. It is a way to take stock of past years and dream of those yet to come."

19

VARROA: A PARASITE MESSENGER

Pubescent humanity, awkward and clumsy,
Stop and breathe, and listen a moment,
This little messenger has lessons to teach.

The Disconcerted Bee

Thursday, 5 July

My young neighbour Zoé, nine years old, is on holiday. She is absolutely fascinated by the bees. "You call me when you are going to the hives," she says to me, "I want to come with you!" I invite her regularly, for the eye of a child always brings me a fresh view of the world.

This morning she meets me near *Demeter*, a beautiful colony installed in a *Warré* hive.[1] The hive is active and serene, and we take up our position close to the entrance. On the ground in front of the hive, a bee is walking a frantic zig-zag. "Look," says Zoé, "this bee looks bizarre! Can I pick her up?"

"Yes, go ahead."

The bee climbs up onto her small hand. I deposit a droplet of honey on her palm and the bee comes to lick it; her stillness allows us to observe her in greater detail. A small brown and oval capsule is riding on her back: a varroa mite. The wings of the bee are slightly deformed, a sign that she is infected with a virus carried by the parasite.[2] "She is sick..." says Zoé, "can we help her?"

Not without some effort, for the mite has ensconced itself in the fold between the thorax and the abdomen of the bee. Armed with a pair of tweezers, I gently extract it. "Maybe she wants to go back into her house now..." murmurs Zoé. She places the disabled bee on the landing board. At once, a guardian detects her and blocks her passage; the infected bee is not allowed to enter the temple. She does not insist, but turns around and, unable to fly, allows herself to fall back down to the ground. "This bee is given over to die, Zoé..." Contrary to the common reaction of other visitors, Zoé does not express any sentiment of sorrow or pity. She accepts death as part of life, without judgement or comment.

Sick bee on Zoe's hand

---

1   This is a vertical *top bar* hive invented by Abbott Warré during the 1920s. The square frames are smaller than those of conventional hives and the honey is harvested from below, which reduces the impact on the colony.

2   DWV = Deformed Wing Virus

"So can I keep it in a box at my house, to observe it?" she asks.

"Yes, you can." I smile.

Co-evolution

"The microbe is nothing, the terrain is everything!" Likely uttered by Antoine Deschamps or Claude Bernard in the late 1800s, they differed in opinion from their more famous contemporary Louis Pasteur, who dedicated his life in combat against the 'wicked microbes'. Deschamps in particular supported the view that every living being exists in permanent interaction with their environment and that their health depends upon their capacity to co-evolve with their surroundings. Let us watch how the varroa mite comes to legitimise this assertion.

One millimetre wide, the varroa is a small mite which subsists on the vital energy of bees. Its original host is *Apis cerana*, an Asian bee. During their co-evolutionary journey, *Apis cerana* learned to control the presence of the parasite by developing adaptive behaviours, so much so that the varroa does not threaten the survival of the species.

With the intensification of agriculture that took place during the mid-20th century, the varroa mite was introduced to Europe. Robust and adaptable, it spread to almost everywhere in the world in a few decades. Today, only some regions of Oceania, central Africa and north America, as well as a few remote areas like the Isle of Ushant, are totally devoid of its presence.

If, thanks to the magic of co-evolution, *Apis cerana* and Varroa have developed a stable and balanced host-parasite relationship, cannot our occidental bee do the same? The inherent logic of Nature allows species to adapt themselves to the challenges offered by life. Unhappily, few colonies have the leisure to attempt it. Indeed, at the global level, the vast majority of hives managed by humans are systemically treated with acaricidic products. Many beekeepers regard treating the hives as indispensable for 'saving the bees', and some accuse those who do not treat of threatening the health of other hives through contagion.

However, interference from the beekeeper, undertaken to 'heal' his hives, delays the process of the species' adaptation, a necessary one for its return to equilibrium. For the bee, the treatments create a dependency upon humanity. They create the illusion of success in the short term, but carry no sustainable solution for the long term. The parasites develop resistance to the treatments and become increasingly vigorous and resilient. Even though they are considered less toxic, so-called 'organic' treatments, when applied systemically, generate similar effects.

Treatments affect the health of the bees along with the health of the humans who handle them. Even though treatments are usually applied outside of the harvest season, the hive is a continuum... and numerous toxins are found in the wax and the honey.

During the first years of my practice, I tried different methods of natural treatments: *thymol* (a phenol extract of the essential oil of thyme), oxalic acid, essential oils... None of them were pleasing to my bees. Even though natural, these products are irritants and – in all logic – the bees responded to them with irritation. So I abandoned all forms of treatment.

There is a widespread infestation of the varroa mite in my region. Numerous colonies suffer from DWV virus and many are 'recycled' every year. However, little by little, some colonies have developed certain 'hygienic behaviours' which maintain the varroa count below a certain level of discomfort. They are learning to groom each other, and to locate and eliminate infected brood. Other colonies swarm repeatedly or replace their queens, both of which interrupt the lifecycle of the mite with a suspension of egg-laying.

During the month of November 2019, I carried out a regular count of the varroas falling from *Demeter*'s nest, a colony that has been prospering in a *Warré* hive for more than five years without receiving any treatment. I observed a relatively feeble number of mites compared to the average and noticed that 90% of the varroa collected had their legs cut off. The bees being unable to overcome the solid shell of the mites, they have found another way to keep them from doing harm: gnawing off their legs.

Today, even though in some countries there are still laws proclaiming anti-varroa treatments to be obligatory, thousands of bee friends refuse to obey them and, through their experience, prove that the beliefs upon which these laws are based derive from a limited view of the world. More and more commercial beekeepers are setting about reducing or ceasing treating, as well. Numerous researchers have witnessed that after a necessary phase for adaptation, mortality rates diminish over the long term.[3]

3　David Heaf, author of *The Bee-friendly Beekeeper*, is one of the forerunners of treatment-free beekeeping. Today, one can find on the internet a multitude of references and studies showing the benefits of abandoning treatments.

The Voice of the Bee

"Varroa is an *agent of decline*. Decline is not to be considered as something wrong or which should not occur. On the contrary, decline is an essential phase of the *life-death-life* cycle. Just as the moon waxes and wanes, Life waxes and wanes.[4]

The agents of decline form a vast team of skilled workers that are indispensable to the functioning of an ecosystem. They are cleaners, and we find among them the decomposers (bacteria, moulds, woodlice, earthworms, ants...) and the vectors of illness (parasites, viruses, bacteria, and fungi). Their role is to destroy obsolete forms and to divide them into fragments so that the pieces can be recycled and re-employed in the fabrication of new forms. Without the presence of these agents of decline, old forms accumulate, stagnate, and block the flow of Life. Thanks to their transformative action, the elemental particles that have been crystallised into form return to the *great cauldron*. The decomposed form becomes dust once again, primary matter, mortar for the generations yet to be born.

Varroa is an agent of the shadow and the shadow is an ally which helps one to stay awake. Too much light erases depth and puts vigilance to sleep; the shadow unceasingly wakes it up. She defines the contours, she sculpts life.

Varroa is a stimulant to evolution. In her current phase of development, the Gaian ecosystem needs parasites and other vectors of illness to find her balance. Without their presence, the evolution of life in matter would be much slower. They encourage the creativity of their hosts, pushing them to invent ever more innovative and sophisticated morphological adaptations and behaviours. They stimulate their resilience, their strength, and their intelligence.

---

4   The term *decline* may be replaced by *involution, degrowth, decomposition, shadow...*

Today, discord in the relationship between the bee and the human has made our physical health fragile. Through a natural feedback mechanism, an S.O.S. was sent into Gaia's vibratory field. In response to this call, Varroa arrived to lend a hand as a cleanser of the physical dimension of our existence. Varroa eliminates feeble colonies through natural selection and frees up space. This weeding out incites the Hive Beings to develop their intelligence and to invent new strategies for evolution. But beyond filling this ecological office, Varroa is also on a mission to arouse the awakening of human consciousness. The extreme infestation of a parasite at the core of a system is an alarm bell, pointing out an imbalance. Through the questioning it provokes, Varroa solicits the re-examination of an obsolete paradigm and the establishment of a new relationship to Earth, and to Life.

As long as colonies continue to be enfeebled by behaviour disconnected from natural law, Varroa will continue to play its role as a cleanser, and we are grateful to it. Varroa invites bees and humans to evolve. When humans interfere in the adaptation process by bringing artificial treatments into hives, evolution is slowed. We need time and space to develop our creativity and to co-evolve with this ally. Co-evolutionary adaptation is a passionate game, one to which we bring ourselves gladly.

Morphic field are fields of probability, and Hive Beings contain the knowledge of all possibilities in their matrix. As soon as a challenge is put forward, the colonies are invited to test new behaviours. As soon as one colony discovers a behaviour which augments her resilience and her vitality, this behaviour is selected and reproduced more and more frequently. Each time that a behavioural pattern is reproduced, its *imprint* is deepened upon the field which carries it. Through force of repetition, these new behaviours become habits, which then circulate in the planetary field through morphic resonance. After a certain period of time, a *critical mass* of colonies adopting this habit is achieved and the entire species can then *download* it naturally, without expending any further effort to learn it."

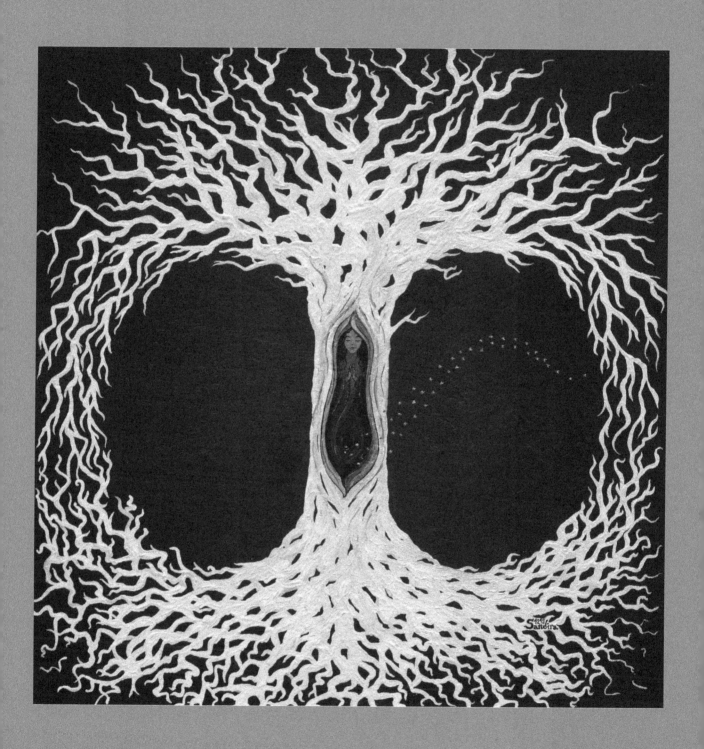

20
THE TREE: SHELTER AND FRIEND

Our wax is your wood,
Our antennae your branches,
Our songs your deep roots.

The Ogress

Sunday, 8 July

This morning, I am creating a space for rest and contemplation at the foot of an old cork oak with a peculiar energy (you might say a kind of ogress with an easy-going gait). A spiral of sound issues from her open mouth, low tones swollen with millenary wisdom.

I bring a stool to explore the cavity and, in her heart, discover a few pieces of old honeycomb, vestiges of a bygone colony of bees. My hand plunges to the bottom, curious, and draws up black gold: fresh humus with a sharp, delicious odour, full of worms and insects.

At the foot of the tree, an enormous dead limb is lodged. Dead and yet alive, its knots and branches, sculpted by the hands of a master, are fascinating. I suspect that its fall, several years ago now, made the colony flee from their home in the heart of the tree. When I set the large branch upright, it turns into a powerful and proud dragon. A little while later, a buzzing sound: a bee flies around the displaced branch with some interest. She lands and forages along the rotten wood that my action has exposed to air. She evidently finds reason for satisfaction from something in this decomposing matter.[1]

Honeybees appeared on the earth more than 100,000,000 years ago. Long before humans put them in boxes, they most often lived in hollow trees (dead or alive). Today, such trees are rare. Considered useless, ugly, vectors of illness, or simply good for firewood, they are eliminated almost systematically.

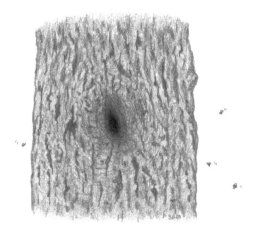

Entrance to a cavity in a chestnut tree

---

1 Bees are nourished by the minerals offered by the mycelium which lives among the roots of the tree or in its decomposing wood. Mycologist Paul Stamets and entomologist Steve Sheppard at the University of Washington have demonstrated the role of certain mushrooms in stimulating the immune systems of bees and their ability to meet the DWV virus. Details of their research are available online at the site *Nature*.

Happily, a few places still exist where colonies have the pleasure of living in trees. Recently, I had the good luck to come across three: the first was in the stump of an old olive tree, the second in the roots of a eucalyptus, and the third in the trunk of a chestnut. I passed long moments at their sides, listening to the Deva speak to me. As colony collapse rages through manufactured apiaries, these 'out of the box' colonies emanate an extraordinary vitality.

The Voice of the Bee

"Bees love trees as much as trees love them. When they share living space, their morphic fields embrace and complement each other. The field of the colony augments the field of the tree, and vice-versa.

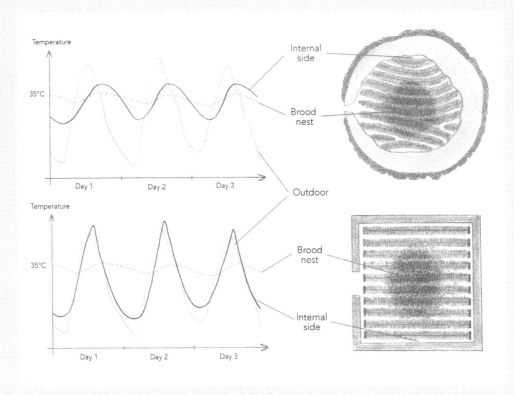

Comparison between the temperature cycles of a natural cavity and those in a conventional square hive

Whether it is living or dead, a tree shelters a vast community that is constituted of numerous inhabitants. Ants, mushrooms, birds, and squirrels live in symbiosis with their host. The tree is a super-organism with a powerful and generous field; each species occupies their specific niche within it. These niches are not isolated from each other; on the contrary, they overlap, permitting the Life energy to pass gradually from one to the other. A clean divide does not exist. Everything is interwoven, everything gets entangled; this intimacy holds the world.

Colonies of *Apis mellifera* who live in the wild are often the issue of swarms escaping from colonies kept by humans. When they have the possibility, swarms which escape from manmade hives like to live in the bosom of their old wooden friends. Their cavities offer round, organic forms. There, our architects build comb in a manner which optimises the circulation of air and humidity, which, in turn, avoids or substantially reduces the development of moulds. The gnarled walls are coated with propolis. We drink the condensed water which trickles down them; clear and pure, it is our medicine. The thickness of the walls provides a high degree of thermal insulation.

In comparison, the corners of conventional square hives are extremely difficult to heat and swell each winter with mould. Pre-built frames impose an inorganic construction which heightens the draughts. And the walls of these hives are so thin that most of our energy is spent maintaining the core of the nest at a suitable temperature."

Wild or Domesticated?

Other than in some remote natural places, honeybee colonies which are wild today are swarms from production hives or their descendants. Does their passage through the system of human exploitation make these bees domesticated animals? This depends, surely, on the definition of the word 'domesticated'. So, once again, the Bee Deva invites us to break down this dichotomy. *Neither wild nor domesticated, the bee is a bridge,* she tells us, *and inspires you to be one as well. We physically arrived on Gaia long before human beings and when you emerged, a mutual curiosity developed. Conscious of our respective potentials, we tamed each other. We offered our honey to you, and our medicine; you offered shelter and gardens to us. However, the contract that was established between our souls involves allowing our wild nature to speak. Our 'rewilding' and your own are indispensable for a true and effective collaboration of our talents. It is essential for you to allow a sufficient number of colonies to live in trees if you would like some harvest hives to survive. The Bee and the Plant can not live without each other. Therefore, dear humans, if you would like to help us, use your creative powers and plant some trees, regenerate the soils, nourish biodiversity, and leave the dead trees alone.*

Rewilding the Bee

In October 2019, my team and I visited a large rewilding project for *Apis mellifera,* underway in the south of Spain by someone truly impassioned about bees, Jonathan Powell.[2] During this memorable encounter, we learned an ingenious technique for increasing the number of colonies occupying hollow trees. In Andalusia, much like in Alentejo, we find many oaks which have cavities with openings that are too large for colonies to comfortably settle themselves. The technique is to use cob mortar (a mixture of clay and straw) to reduce the entrance of the cavity to a suitable size.

The fall of that branch from 'ogress', the large oak in the story I recounted above, rendered the tree inadequate for the installation of a new colony. Upon our return, inspired by our Andalusian experience, we created our first 'cob cavity' there. To make the mixture waterproof, the cob was embellished with cactus juice, then covered with a final coat based on rye flour and linseed oil. We added powdered shungite, a stone offering exceptional protection from electromagnetic radiation.[3]

The following spring, we installed a swarm there which was collected from a neighbour. From all appearances, the old ogress delights in being inhabited by this vital young colony. During our courses and open days, we invite many groups to stretch out around this century-old tree. Guided by the sound of the drum, they are encouraged to imagine themselves in the hollow of the ogress. The experience has been profoundly moving for many participants. "I never thought to visit inside a hive with my thoughts. The images were clear and powerful. The bees welcomed me with benevolence. The odours, sounds, and architecture of the combs, it was magnificent… I was intoxicated. I felt the vitality, the anchor, and the security that the tree gives to the colony," my friend Uri confides to me after this experience, stars in his eyes.

---

2   A project on eco-farm La Donaira, an estate of 700 hectares in Andalusia, carried forward by Jonathan Powell, administrator for the organisation *Natural Beekeeping Trust* (based in the south of England).

3   The pectin contained in the juice of Barbary figs (*opuntia ficus-indica*) acts like a glue which makes the cob mixture waterproof. Shungite is a mineral of organic origins mainly from Russia; it offers extraordinary qualities for the protection of bees (and humans) from electromagnetic radiation. The noxious waves and toxins emitted by human technologies alter the dynamic of morphic field rotation, and the unique molecular structure of Shungite has the property of re-establishing a biocompatible rotation.

The Voice of the Bee

"Trees are an essential organ of the Gaia organism; they are antennae, receivers and broadcasters of vital information. Through their branches and their roots, they channel the information flow emitted by the stars, the subterranean entities and other guardians of Gaia. They harmonise the cosmic and telluric currents which give our planet her exceptional vitality and biodiversity. Along with the movement of their sap, their respiration behaves like a engine that pumps and regenerates the blood of the earth. They interconnect the vertical and horizontal energy flows and thereby regulate climates. Old trees, whether they are living or dead, are among the guardians of planetary wisdom.

The foragers possess an inner compass aligned with terrestrial fields which automatically allows them to orient themselves and find the path back to their hive. This magnetoreception system is affected by electromagnetic frequencies that are out of alignment with natural law, such as those broadcast by new human technologies like, for example, 4G and 5G. Disoriented, the foragers get lost, the colonies are thinned out and disappear. These *antibiotic* frequencies increase, among other things, the rate of free radicals (positive ions) in the atmosphere.

Trees filter and neutralise these radiations. When a colony lives in a tree, she profits from the cloud of negative ions which the tree emits around itself. These anions are necessary for the proper functioning of all living beings, bees as much as humans. They permit the oxygenation of tissue, boost the immune system, and reduce vulnerability to stress.

Gaia is an enormous reservoir of negative ions; they are transported by tree sap and distributed in the air. The size and height of each tree create differing electrical potentials, which also generate the formation of anions. The anions bring prâna into the air and clean the atmosphere of bacteria, viruses, and pathogenic moulds.

When the possibility is available, Hive Beings value dwelling places which are situated several metres off the ground. Telluric networks (such as the Hartman network) progressively dissipate as they move out from the physical surface of the earth. The energetic lightness of the earth body's higher strata facilitates the revitalisation work of the ethers undertaken by the colony."

Hives of Biodiversity

A raised trunk-hive

Not everyone is lucky enough to have hollow trees close to home, and there are numerous alternatives to offer bees which recall their natural habitat. The drawing above illustrates one model of a trunk-hive, very much appreciated by the bees, that our network is distributing now in Portugal and England. Its thick walls largely reduce temperature fluctuations inside the nest; the conserved energy is invested in strengthening their immune system, the development of adaptive hygienic behaviours, and the fundamental energetic work which they offer to the planet.

I am also experimenting with various methods of improving the frame hive; the *symbiotic floor*, for example, is an idea which is beginning to prove itself. A box filled with decomposing organic matter is attached below the nest where, like the bottom of a hollow tree, the litter creates a buffer zone that is favourable to synergetic interactions. The bees are nourished by certain substances produced by the mushrooms and micro-organisms which live in this layer of humus.

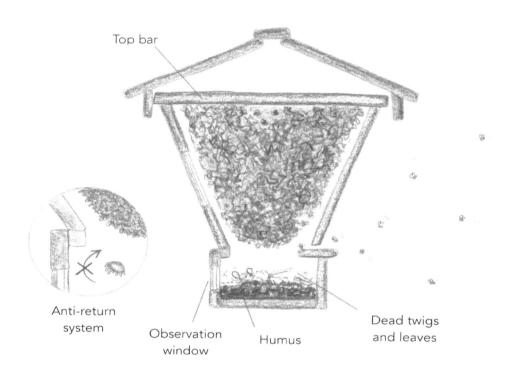

Top bar

Anti-return
system

Observation
window

Humus

Dead twigs
and leaves

The symbiotic floor

This system offers an interesting alternative to mechanical methods for reducing the presence of varroa. Since the box is slightly larger than the bottom of the hive, mites that come off their hosts fall into the litter, and they cannot climb back up the walls to reinfect the nest. They end their lives decomposed by the micro-organisms present in the humus and, slowly, become dust once again…

21

DEATH AND THE BEE

Down here on this stage, in a world seemingly solid,
Life dances with death, all whirling and smiles,
Two-sided continuum, lemniscate of Infinity.

Regenerative Death

In 2012, when I was part of the ecology team at Tamera, I had the opportunity to make a dream come true, one which had lain dormant within me for quite a few years: the creation of a *bee garden*. This garden is a haven of peace, a downy cocoon in the shape of an egg about 30m by 20m that, today, is plentiful with insects and melliferous plants. Before its creation, the area was an arid desert studded with the skeletons of a few dead trees and, every time I visited, the area induced a distinct constriction of my solar plexus. However, the message from the Bee Deva was clear: this was the right place to establish this bee sanctuary. I received the support of umpteen friends and volunteers along the way: the creation of high protective berms outlining its ovoid form, the planting of abundant trees and aromatic plants, and the installation of two great standing stones – the *antennae* of the garden. The first year, two colonies of bees were installed.

Thursday, 22 September

Autumn equinox and the light is soft, an enchantress. A perfect atmosphere for this portal-day, the pivot between summer and winter, light and dark, life and death... I arrive at the bee garden, greeted by golden sunbeams which slide between big grey clouds. I savour a joy as sweet as silence.

However, a wave of sorrow crashes through me when I realize that *Dana* is dead. Although numerous bees are coming and going from her entrance, I see from the way they are flying that they are neighbours who have come to salvage the rest of the honey. The death of Dana takes me by surprise; engorged with honey, she seemed so strong!

Immediately, guilt assaults me: *What did I do wrong?* I stretch out in front of the hive, connect to the Earth and to the Hive Being of Dana. *Why?* She does not take long to respond to me:

"You did not do anything wrong, Sandira. Death is not 'something wrong'. Wrong is a human invention which Nature does not know.

– But you were so full of life... What made you depart?

– We decided to die.

– Why was that?

– For us bees, death is an integral part of Life. Life and death are two sides of the same coin, the two loops of the lemniscate. Life and death are the day and the night of a great Journey, that of Life with a capital 'L'.

During this last life, we accomplished an important work by cleansing and healing this place. Many memories of suffering were stuck here and several lost souls were pinioned in the dead trees. We are healers, guardians of the threshold; it is our task to free these outworn energies. This work is necessary and we perform it with complete devotion.

This work of liberation requires a lot of energy. Our energetic reservoir being limited during our passage through matter, it is necessary for us to die regularly in order to 'fill up'. So we die with joy, since we know that the other side is magnificent... And we will return soon, don't worry."

I quiver with admiration, even reverence. I, who know nothing about death, who am I to judge it just or unjust? A profound feeling of humility embraces me.

New Perspective on the Disappearance of Bees

This story is followed by many similar experiences. Each year, several colonies die and each time that I react emotionally, the Bee Deva invites me to see the experience without drama and to widen the spectrum of my perspective.

To *de-dramatise* - this does not mean to deny. It is essential to be aware of the imbalanced, fragile state of the bees and the planet today. The invitation of the Bee Deva is about recognising the situation and accepting it for what it is. *To accept does not mean to approve,* she says to me, *but it is to admit, without judgement, that the situation exists and that we will need to deal with it. Acceptance is like a base from which one acts in a constructive manner. Then it becomes possible to use creativity to put forward something new, which naturally brings about the dissolution of that situation which has become obsolete.*

Over the course of my years of research and of conversations with the Deva, my perspective on the disappearance of the bees has fundamentally transformed. Like many others, I started beekeeping with a focus on 'what doesn't work'. I denounced 'irresponsible' apiculture practices and uttered guilt-inducing words. Without realising it, I was replacing one dogma with another and stagnating in a dynamic based on conflict. While I wanted to 'do good', I was nourishing the *pain body* with which humanity has garbed the bee.

The bee does not know suffering. The bee adapts; she comes to terms with present conditions and if she needs to pass through death to continue to advance, this is not a problem for her. That which we regard as 'the bee problem', is, in reality, our problem.

Integrating this perspective on the world was a great inner revolution for me – a sort of *little death...* As Eckhart Tolle says, "The secret of life is to die before dying and to discover that death does not exist."

The Voice of the Bee

"Death is only hostile when it is misunderstood. To love life, one must love death. The passage through death cleanses and regenerates. Death is as lovely as the velvety voice of autumn which calms and tempers the passions of summer.

Human religions over the last two thousand years have wrapped death in a veil of fear and suffering. However, death is neither good nor evil; only your thoughts render it so.

Life with a capital 'L' is a lemniscate. Thus, life (small 'l') and death are two complementary aspects and inseparable from Life.

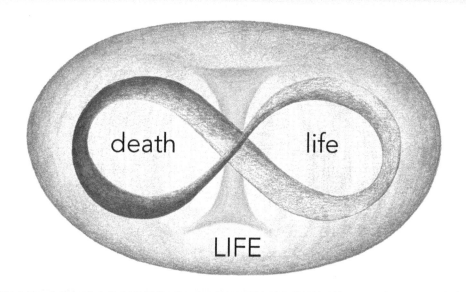

The Life-Death-Life cycle

The *life – death – life* cycle is the engine of the world. Life and death are each as fascinating as the other. The journey through life is a school of learning, where the Being experiences the curious marvel of density. The journey on the other side draws near to Source; it enables a recharging of batteries, as well as interactions with the material world in an etheric form. These two interconnected adventures follow each other – or *superimpose* upon each other – in a continuum.

The massive disappearance of bees in numerous regions of Gaia reveals many messages to humans who know how to listen. To understand them, you must learn to overcome the wave of emotion which these departures initially provoke, and to broaden your spectrum of perception.

Understand, dear humans, that in some areas, there are simply *too many* honeybees. This declaration may seem shocking for those apiphiles who seek to protect them and worry about their decline. And, nevertheless, it is true: in many places around the world, the number of colonies 'in boxes' is too high. Already weakened and impoverished by human practices, many ecosystems suffocate when they are occupied by millions of bees who are programmed to produce honey in order to fill boxes that are incessantly emptied.

There is a vast team of pollinating insects whose efficacy depends upon the diversity of its members. If the number of honeybees goes above a certain percentage, other species decline from a lack of resources. Solitary bees and bumblebees are indispensable actors without whom the play cannot go on. Our work is an orchestra, not a solo performance. An apiarian monoculture will never bring the land to flower.

The fact that a great portion of the harvested nectar is extracted from the system by beekeepers creates a major leak of energy which further aggravates the imbalance. The hive beings can compensate for this disequilibrium for some time, but beyond a certain threshold, Gaia unleashes her regulatory systems. The agents of decline, such as Varroa and the Asian hornet, are sent to recalibrate the flow of forms. They help us to cleanse and regain our inherent power. Sometimes entire apiaries disappear from one day to the next, leaving behind only their carcasses of wax and honey.

The journey 'through the other side' allows Hive Beings to observe the incarnated world from a distance and to introduce strategies for their next passage through matter. Temporarily freed from material weight, they cleanse and regenerate themselves. Today, the genetic palette of our species is debilitated. Conserving the genetic information of weak lineages is not desirable, and death acts as a sieve which preserves the strong information and eliminates the detrimental.

If out-of-tune human practices have eroded the robustness of our physical envelopes, our Essence is not affected. Our Soul is pure and untouchable, and our passage through death allows us to reconnect with it. But if the physical form of the bee is disappearing now in a massive way, this is not irreversible. Should the human species open its eyes and take up its responsibilities as a guardian of the planet, a multitude of Hive Beings are ready to come back into incarnation, powerful and revitalised."

Soul Guides

For your power, o death, is great and admirable!
What was, is re-made;
everything flows likes water,
And nothing under heaven will be seen again.
Each form changes into an other, renewed,
And that renewed is called living, by the world,
And death, when that form into another goes.

Pierre de Ronsard (1524 – 1585)

In some traditions, it was customary to tell the bees about a death or a birth in the family. Sometimes a black veil would be lain on the hive for a few days after the death of the beekeeper. In Eastern Europe, bees were frequently engraved on tombs, accompanying the deceased on their journey.

The Voice of the Bee

"Most humans have forgotten the histories which precede their current incarnation. The before and the after, these unknowns, terrify you... For us, death is as much a friend to us as life. We offer to help you reconcile yourselves with the beyond, so you can receive its sacred beauty into yourselves. Beneath her mask of beliefs, Death is an infinite kindness. You have disguised her as a cruel and ruthless reaper and as long as these thoughts continue to dress her up, she will play that game. But when you change perspective and integrate her into the great cycle of Life, death becomes an ally, a companion, a friend.

The Bee Deva is a threshold guardian, a guide for souls. When a human being is born or dies, we offer our services to render the journey smooth and pleasant for them. We know the itinerary and the pitfalls of the road, and do our best to guide the Soul on their way. The passage through death is an adventure, an experience to welcome. And even in the belly of death lay waiting the sparks of rebirth, which shine like stars in the heart of night."

22

CO-CREATION

Bees and humans, gardening the planet,
Acting together, with singular mind,
Building today our world of tomorrow.

Layers of Perception

Monday, 29 October

A windy and confused morning; an ache in my limbs and in my head.

I go outside for a walk in the *montado*.[1] I am absent-minded, absorbed in my mental chatter and chewing over this or that conversation or even what I am going to eat upon return… Suddenly, the warm voice of the Bee Deva resonates: *What you see depends upon how you look*.

Mmmm, this is not the first time she has offered this phrase to me for meditation… I stop, and *look*. It's interesting - although I thought I knew all the nooks and crannies of this side of the hill, I find myself in a little hollow of a valley which, I realise, I have never previously entered. There is a misty, unusual atmosphere; the silence is heavy and as thick as butter. It seems deserted to me. I hold my breath, my antennae on alert.

A sudden buzzing to my left: a dainty forager, I see her on the inconspicuous flowers of a *trovisco*.[2] The next moment, there is another a bit further away, on the small bell-flowers of an arbutus. Then another, and another again. The deserted valley is suddenly overflowing with life! It was all there already – only my gaze had changed. *The observer creates their reality*, the Deva tells me.

Then something happens which is even more strange: all of a sudden, certain elements in the landscape appear particularly bright to me, while the rest become blurry, so that the extremities of the plants stand out from a pixelated background. My system of perception transforms into '*Gimp* software', capable of rendering certain layers visible while masking others.

Amused by this game of selective perception, I *ask* the programme to conceal the vegetal layer and to keep only the bees. A map of the collection of foragers forms in my brain; their itineraries are added in a coloured outline. The image which appears is a mosaic of geometric motifs reminiscent of the flower-of-life. I close my eyes and enlarge the viewing angle: the map stretches to cover the region, to the country, then to the entire planet…. All the land of Gaia is clothed in this golden weave. Magnificent, the image is engraved in my memory as though illuminated.

---

1   The characteristic landscape of Alentejo, the *montado* is an ecosystem of pastures with sparse undergrowth (mainly cork oaks).

2   The *trovisco* or flax-leaved daphne (Lat: *Daphne gnidium*) is a shrub whose small flowers emit a sweet and captivating odour.

Homage to the Human Being

In *alternative* or *enviro* circles, there is a tendency to speak of the human being with a certain bitter considering-all-the-harm-done-to-the-planet attitude, along with shouldering a goodly amount of guilt and sighing with chagrin. The Bee Deva smiles at these scenes. Indeed, while it is certainly important to realise the degree of immaturity of some human behaviours and their consequences for the health of Gaia, she invites us to orient our attention towards the regenerative spirals of the earth system. *Energy follows thought*, she tells us, *and when you focus on the pain body generated by Humanity, you feed it with your attention. Your identification with the pain body gives it even more strength and energy than it already has.*

Today, we are bidden to revolutionise our perspective and to screw a wide-angle lens on our inner camera. A great call for planetary cooperation is ringing all over the world. The transition from the Piscean Age to the Age of Aquarius is characterised by the passage from individualism to cooperation and interdependence. The 'I' becomes 'we', even while valuing the unique qualities of each individual. *Your mode of connecting with the world changes the experience that you create*, adds the Deva. *Dare to incarnate your profound Being and be the change that you would like to see in the world! Thus, you honour the gift of your human incarnation.*

Bees love humans. Not our whims, nor our oversized Egos, but our Essence. Thus, to encourage us to anchor ourselves on the constructive side of Life, the Bee Deva offers us this homage.

The Voice of the Bee

"Soul friend, human soul, know that the Human Being fascinates us and inspires us, a veritable poem... Your physical body is a treasure of creation. So many tools available, endowed with such precision! With only two feet and two legs, you walk, move, climb, jump; with your capable hands you catch, carry, caress, feel, transform, create... You have at your disposal a myriad of sense organs to perceive the tangible and subtle worlds, musical vocal cords to express the power of the word and to sing out with sound. And, centrepiece of the painting: your incredibly sophisticated brain that makes you understand and think about the world.

You have no physical wings, but your heart has wings. The *field* of your heart is without limit, and when you activate it, the love which is released is beyond all measure. Its vibrations are a *medicine*, a curative balm which heals wounds. When you take the time to offer this energy to us, we drink it like fresh spring water.

Your will is powerful; it pierces through veils. It is able to guide the most delirious ideas towards their manifestation. The human will is a double-edged sword which you must learn to master and to handle with care. It gives you the *choice* to serve violence, deceit, and theft, or to serve the Great Work of Creation. The human ego can be terribly destructive when it takes control, or the bearer of an awe-inspiring potential when it is placed in service to the Soul. Humanity is in the process of acquiring a new maturity, and we trust you. A great surge of awareness and empowerment is at work in the human species; it is a joy to witness it.

To conclude this homage to your Being, dear humans, know that that which we admire above all in your incarnated Being is your inner artist, your creative spirit, this capacity to manifest divine grace through form. Whether it is through music or dance, painting or writing, philosophy or science, the creation of landscapes or extraordinary dwellings and flowered gardens, you know how to reveal the invisible within the visible, to channel the abundance of Source, to use the body to sculpt beauty, and to give form to the sacred. And this is only beginning. We are at the dawn of a New Earth, a new paradigm, motivated by cooperation and guided by the desire to serve the manifestation of a *planetary garden*. Wise women, wise men, let yourselves be led by your true Essence; guide your ego to let go, and realise the immensity of your potential."

Love and Vision

Sometimes it is difficult to receive compliments. This eulogy to the Human Being touches my heart and loosens an avalanche of questions... Why, then, so much violence, separation and suffering? Why is there war and the reckless exploitation of the resources of our planet?

It seems to me that humanity is mired in a sea of limiting and debilitating beliefs.... Can we really get out of this? Waves of doubt regularly assail me. Can I really overcome them? What is powerful enough to seduce the human to put their potential in service to planetary cooperation? The Deva replies...

The Voice of the Bee

"Love and Vision.

Vision gives direction to Life. It is composed of dreams which aspire to the new and constantly invite you to move forward. It gives direction and provokes movement, transformation, evolution. As for Love, this is the cohering force which links all Beings. It brings about the manifestation of Vision in form.

Violent behaviours, motivated by that which their authors believe to be love, are always born from an attachment to form. This possessiveness, this fear of emptiness, is engendered by an illusion – that of the severed umbilical cord, a separation from Source. The more the human being believes themselves to be separate, the more it will manifest in reality. This separation is the myth to dissolve in order to live in peace.

The beliefs that are the basis for this myth have deep roots, anchored in the human psyche. However, as solid as it may seem, no belief is immutable. Remember, the belief that the earth was flat was an unshakable truth for 15th C Westerners. A new perspective on the world appeared, and here it is – dissolved. It is the same for all dogma and law, including the physical laws of Nature."

Free Energy

The social programme for the majority of human beings is based on the fact that all material things will, one day, decline. We build our lives upon this axiom. Our planetary economic system is founded on scarcity and the quest to possess it. The material resources of the planet are limited and those who have access to them are rich, while the others are poor. So, the materialist vision of the world condemns us to the eternal fear of lack, of emptiness… of death.

However, today quantum physics demonstrates this strange fact: protons do not die. This primal matter at the foundation of all form shows no sign of decline. Or, to be more precise, it appears that protons pass their time by dying and regenerating without end, thousands of times per second. The energy which fuels this infinite dance of life-death-life is inexhaustible.

All forms are made of protons, assembled through a sublime cosmometry. Each one of our cells contains thousands of protons which permanently regenerate themselves, maintained in life by this limitless source of energy. This holy grail, this mythical fountain of youth for which many explorers sought in the exterior, is found within each of our atoms.

More than ever before in our History, today the Devas invite us to learn to draw upon this free energy and to channel it. Thousands of pioneers, seekers of truth and alternatives to our failing system, are discovering the art of receiving, using, and guiding these frequencies in multiple domains. Breatharians draw their nourishment from there, psychics draw their intuition from there, artists draw their inspiration from there, and quantum mechanics draw fuel from there for the machines of tomorrow.

A profound revolution in science is underway. I feel my trust affirmed when I hear Frank Wilczek, nominated for the 2004 Nobel prize in physics, openly declare that 'we are children of the Ethers'.[3] With humour and simplicity, he reminds us that before becoming matter, we are pure energy. A century after the precious discoveries of Einstein, the scientific community is beginning to take off the blinders and open to the possibility of developing technology which respects the Earth.

---

3   Citation from the conference *Materiality of a Vacuum: Late Night Thoughts of a Physicist*, 2017, available online.

The Bee Deva tells us: *Each species has a particular gift to offer the earthly community. During the last two centuries, thanks to their intelligence, humanity has learned to play with electromagnetic fields and to use fossil fuels to operate their technology. This phase, necessary up to a certain point, has been costly to the planet and all her inhabitants. Continuing down this road will lead to a dead end. Our spatio-temporal position at the core of the Gaian system gives you a fundamental choice: either follow the involutionary curve or take the decision to place your genius in service to All.*

To access this unlimited source of energy, adds the Deva, *it is necessary to understand that, fundamentally, we swim in a world of abundance, despite all appearances.* This is easy to say, but how can we make it happen, when we are entangled in a system of limitations from birth? Today, many tools and techniques demonstrate their effectiveness in supporting the deprogramming and reprogramming of our belief systems and the conscious reconnection with universal energy fields.[4]

The Voice of the Bee

"The more you understand how universal laws function and develop the capacities of your conscious mind, the more you are able to direct the lines of manifestation in the world. The art of conscious intentionality is one of the most powerful capacities at the disposal of the human being. Bees don't have the level of individual consciousness needed to practise this art consciously. However, we hold fundamental knowledge about the functioning of the Universe, its geometry and its language, which can help you develop your capacities with accuracy and grandeur. We are disposed to share our knowledge with whoever would like to receive it. Bees, nature spirits, and other guardians of the planet are able to make many things which are inaccessible to you without our assistance.

Information in→forms, it creates form. The art of intention consists of combining the activation of your various bodies with the goal of manifesting forms or events. At the core of the infinite field of potential, a probability is selected and formulated by the mental body, magnetised by the emotional body, and then coagulated in the physical world.

---

4   The tool that has been the most efficient for me is called the Lightning Process, developed by Phil Parker, an English osteopath. Three-day seminars in various languages are available to learn this simple technique.

By applying this process in a concentrated and assiduous manner, it is possible, for example, to modify locally the gravitational field in order to move allegedly immovable objects. Thus were moved bricks of many tons, to make the Egyptian pyramids, and certain Incan walls. If a group of humans, accompanied by allies in the subtle worlds, gathers and emits a collective intention that is as powerful and as pure as crystal, form will allow itself to be sculpted, as malleable as warm wax."

Network of Networks

The garden cannot be taught, it is the teacher.
Gilles Clément, *Jardins, paysages, et génie naturel*

When they learn that I work with bees, many people show distress and say something like "The bees are in trouble, we must save them." In response to this naive and touching injunction, the Deva says: *Bees don't need to be saved. If there is anything to heal, it is the relationship which you maintain with the Bees and with Nature; this is where the source of the imbalance resides. To change the world, begin with changing yourselves, and observe the resonance that this generates around you. Reconsider your belief system and adapt it to your new discoveries. Train your ego to be in service to Soul. Create networks, and gather yourselves! Bring together your respective qualities for seeding new fields of consciousness, and plant forests and gardens to bring new life to dried fields – those of your heart as well as those of the planet.*

Indeed, the beauty of our era is expressed in the dance which unites the local and the global. Communities of all kinds are forming around the world, like thousands of interconnected acupuncture points, in a formidable network of meridians. Research groups gather locally and internationally under the form of eco-villages, amateur and professional networks, talking circles, and many other initiatives. The revolution of our communication system – and, notably, the development of the internet – renders these gatherings easier and easier, and more accessible.

The networks of Gaia

In the spring of 2017, a few friends and I launched the *Bee Wisdom* network. Its objective is to connect the people and places which give their attention to improving the relationship between humans and bees, to the mutual learning they can bring, and to the co-creation of a planetary garden.

One of our research group's experiments is to place our hives in an energetic network, locally and globally. At the local level, the hives are linked to a focal point (an erect stone, for example) and form an energetic grid. All the local grids are then interconnected by their focal points. The Bee Deva suggests that this interconnection accelerates the circulation of information from colony to colony, through morphic resonance. New adaptive behaviours developed by the bees faced with the challenge of varroa or the Asian hornet, for example, thereby spread more quickly through the network. Feedback loops are boosted and the more fragile members of the system receive the support of those who are more vigorous.[5]

---

5 To create a local grid of colonies and connect it with our network, visit our internet site: beewisdom. earth.

The Voice of the Bee

"The guiding plan of the earthly community consists of creating *a planetary garden*. In the noble and large sense of the term, a garden is a cathedral. The planetary garden is a living cathedral in which each inhabitant consciously evolves and unfolds, holding the hand of their neighbour. A world where marvels have their place and are discovered at every moment. A world where the more we give, the more we receive; the more generous we are, the richer we are. The planetary garden is a co-created and co-creative space, where each member is an active and responsible participant.

Dear humans, it is time to fully shoulder your responsibilities as gardeners of the planet. This is not to create a sterile little garden, 'all pretty and clean'. This is not to control everything that happens nor to impose your rules and constraints. Quite the contrary. The role of a true gardener is to listen to each member of the team: from the earthworms to the elephants, from a blade of grass to a baobab. The gardener is attentive to the whispers of the wind, to the influence of the moon, and to the messages in fairy song.

The real world is the result of a collective dream which the ancestors dreamed. To create the planetary garden, dream it! You are the weavers of fields; choose the fibres and motifs with care. Gaia offers us her body to touch, to sculpt, to render sublime; let us honour it."

CONCLUSION

Here we are, at the end of our journey. The end… or the beginning! I hope that the messages delivered in this book continue to accompany your *apiphilic* peregrinations. It has been an honour for me to share them with you.

The writing and illustration of this book has been a true initiatory journey. Two years ago, I weighed 33 kilograms and moved with difficulty. My skeleton-like body starved, but did not digest physical nourishment properly, no matter what type of diet I offered. Even while some of my nearest and dearest doubted it, I have always cultivated a profound desire to live. Despite pain and frustration, I received the trial of illness as a gift which Life offered to help me grow. In October 2018, I barely achieved 36 kilos and, every winter, I shivered as soon as I emerged from my blankets, no matter how many layers of wool I wore. Although endowed with a profound *joie de vivre*, I was fatigued by this complicated life. That is when the adventure of this book began, fifteen months ago today. I smile from the synchronicity; many of my paintings have 15 bees, of which the essential root is 6 (1+5), the symbolic number of the bee.

Fibromyalgia is a very curious 'disease', one which cannot be understood when approached through the purely rational window of the mind. Like many other syndromes which conventional medicine doesn't really know how to handle, fibromyalgia is an invitation, a profound appeal, for the soul to change its relationship with the world. Only a holistic approach towards existence can guide its resolution.

The publication of this work is a childbirth. The healing power which accompanies its gestation and its birth is extraordinary. During the first two months of pregnancy, I consecrated eight to ten hours per day to writing, awakening before dawn and most of the time sitting in my bed, legs stretched out, in the least uncomfortable position for my sore body. A beeswax candle and a propolis diffuser invited my devic muse to my side. At the end of these two months, I had gained 10 kilos and the first draft of the manuscript was ready. At this stage, I imagined that the work would be edited several weeks later. Today, I smile at my naïvéte; to be born, the baby really needed refinement. It is during this year of maturation that, step by step, guided by the devic wisdom contained in these lines, my physical body recovered its power, its vitality, and its ideal weight.

At the hour of writing these last lines, I feel full of gratitude for this voyage, so rich and revelatory. The flame of my candle flickers, a soft light against the veils of dawn; I close my eyes. Serene and warm, the voice of the Deva resonates: *Soul friends, human souls, dare to assert your grandeur with humility and recognise, at every instant, the miracle of being alive. Allow yourselves to be guided by the wings of your soul. Seed your flowers, offer your fruit, and invite the bees to pollinate your dreams. The source of the planetary garden dwells within you.*

Sandira Belia

Sunday, 26 January 2020

Serra da Estrella, Beira Alta, Portugal

BIBLIOGRAPHY

My Preferred

■ Jacqueline FREEMAN, *Song of Increase: Listening to the Wisdom of Honeybees for Kinder Beekeeping and a Better World*, 2nd ed., SoundsTrue, 2016.

> I devoured this book when I discovered it in 2015. I realised that I am not alone in communicating with the Bee Deva and that her wisdom is worth sharing. On one hand, Jacqueline shares her discoveries, her experiences, and her questioning with humility and simplicity. On the other, she offers a series of messages received from the Bee Deva which combine and overlap with those presented here. Reading *Song of Increase* greatly inspired me to give birth to this work.

■ Simon BUXTON, *The Shamanic Way of the Bee: Ancient Wisdom and Healing Practices of the Bee Masters*, 2nd ed., Destiny Books, 2006.

> A captivating account for lovers of initiation stories. Written with grace and poetry, it tells of the author's adventures during his journey learning the shamanic wisdom of the bee from Bridge, master beekeeper with the Path of Pollen. The Sacred Trust, co-founded in the south of England by the author, offers seminars delivering the teachings of the Path of Pollen (today known as the Lyceum), as well as many other teachings. Strongly recommended for those who wish to deepen their exploration.

■ Eric TOURNERET, Sylla DE SAINT PIERRE, Jürgen TAUTZ, *Le Génie des abeilles*, Éd. Hozhoni, 2017 (currently available only in French).

> This magnificent work is simultaneously a treasure for the eyes from the beauty and quality of its photos, and a mine of the most recent information about the biology and the ethology of the bee.

- Sue MONK KIDD, *The Secret Life of Bees*, Tinder Press, 2003.

  This passionate novel tells of the encounter between Lyly, 14 years old, and three extraordinary apiculturists in South Carolina during the 1960s. The delicious style and depth of the narrative rocked my world.

- Rose-Lynn FISHER, *BEE*, Princeton Architectural Press, 2010.

  A book of extraordinary photos taken with an electronic microscope. In it, we discover the incredible engineering of bee organs such as the multi-faceted surface of her eyes or the mini-hooks which allow them to join up their wings. This microcosmic dive is breathtaking.

Other Works I Value

Bee Biology and Ethology

- Jürgen TAUTZ, *The Buzz About Bees: Biology of a Superorganism*, Spring, 2008. Original title *Phänomen Honigbiene,* 2007.

  Superb photos from Helga R. Heilmann. The content is dense, reserved for enthusiasts.

- Thomas SEELY, *Honeybee Democracy*, Princeton University Press, 2010.

  Very technical and scientific, excellent for those who are passionate.

- Rudolf STEINER, *Nine Lectures on Bees: Given In 1923 To The Workmen At The Goetheanum*, Northern Bee Books, 2020.

  Some may find it dated, but this book includes beautiful pearls that are still very relevant today.

- Torben SCHIFFER, *Bee Evolution* (forthcoming).
  Original title *Evolution der Bienenhaltung : Artenschutz für Honigbienen,* 2020.

  I had the opportunity to meet Torben at the 2019 *Learning from the Bees* conference in Berlin. His passion and love for the bees shines. I have not yet read his book, but I foresee that his work reveals research of major importance on the true nature of the bees.

Alternative Beekeeping

▪ Heinrich STORCH, *At the Hive Entrance*, European Apicultural Editions, 1985. Original title *Am Flugloch,* 1951; PDF available for download at the Internet Archive.

> This pamphlet is a good source of information for those who would like to minimise hive openings and learn to read what is happening inside the hive from attentive observations of the entryway.

▪ David HEAF, *Natural Beekeeping with the Warre Hive: A Manual*, Northern Bee Books, 2013.

> A reference book which includes an important basis for reflection on natural apiculture.

▪ Fedor LAZUTIN, *Keeping Bees with a Smile: a vision and practice of natural apiculture,* New Society Publishers, 2020.

> I had the good fortune to meet the author (now deceased), who lived not far from the Arctic Circle in Russia. He describes his loving approach to beekeeping with gusto, along with, among other things, the thermodynamics of the winter cluster.

▪ Jacqueline FREEMAN & Susan KNILANS, *What Bees Want: Beekeeping as Nature Intended,* Countryman Press, coming soon, 2022.

Health

▪ Amber ROSE, *Pioneers Companion Workbook, Acupuncture Treatment Plans and Pathways*, CreateSpace, 2015.

> For those who would like to deepen their usage of bee venom in apipuncture, this book offers detailed lists of meridians and therapeutic indications for each point.

▪ Roch DOMEREGO, *The Healing Bee,* Northern Bee Books, 2016.

> This book gives many interesting details on the therapeutic uses of hive products.

▪ Claudette RAYNAL-CARTABAS, *Guérir avec les abeilles*, Éd. Guy Trédaniel, 2009 (available only in French).

> I like this book very much; it links apitherapy with traditional Chinese medicine.

Philosophy and Spirituality

■ Horst KORNBERGER, *Global Hive: What the Bee Crisis Teaches Us About Building a Sustainable World*, Floris Books, 2019.

An artist's vision, for whom the frames of conventional hives are analogous to the mental frames of our society. A poignant call to good will.

■ Mark L. WINSTON, *Bee Time, lessons from the hive,* Harvard University Press, 2014.

Interesting explorations into relationships between bees and societies.

■ Pierre-Henri TAVOILLOT and François TAVOILLOT, *L'abeille (et le) philosophe, Étonnant voyage dans la ruche des sages,* Éd. Odile Jacob, 2015 (available only in French and Spanish).

This book contains pearls and anecdotes about the symbolism of the bee throughout human history.

Lightning Source UK Ltd.
Milton Keynes UK
UKHW051914090223
416667UK00005B/98